Surveys of Modern Mathematics
Volume VIII

Lie-Bäcklund-Darboux Transformations

Y. Charles Li
Department of Mathematics, University of Missouri

Artyom Yurov
*Department of Theoretical Physics,
Kaliningrad State University, Kaliningrad, Russia*

International Press
www.intlpress.com

高等教育出版社
HIGHER EDUCATION PRESS

Surveys of Modern Mathematics, Volume VIII
Lie-Bäcklund-Darboux Transformations

Y. Charles Li
Department of Mathematics, University of Missouri
Artyom Yurov
Department of Theoretical Physics, Kaliningrad State University, Kaliningrad, Russia

ISBN: 978-1-57146-288-6

Printed in the United States of America.

18 17 16 15 14   1 2 3 4 5 6 7 8 9

SURVEYS OF MODERN MATHEMATICS

Preface

One of the mathematical miracles of the 20th century was the discovery of a group of nonlinear wave equations being integrable. These integrable systems are the infinite dimensional counterpart of the finite dimensional integrable Hamiltonian systems of classical mechanics. Icons of integrable systems are the KdV equation, sine-Gordon equation, nonlinear Schrödinger equation etc. The beauty of the integrable theory is reflected by the explicit formulas of nontrivial solutions to the integrable systems. These explicit solutions bear the iconic names of soliton, multi-soliton, breather, quasi-periodic orbit, homoclinic orbit (the focus of this book) etc. There are several ways now available for obtaining these explicit solutions: Bäcklund transformation, Darboux transformation, and inverse scattering transform. The clear connection among these transforms is still an open question although they are certainly closely related. These transformations can be regarded as the counterpart of the canonical transformation of the finite dimensional integrable Hamiltonian system. Bäcklund transformation originated from a quest for Lie's second type invariant transformation rather than his tangent transformation. That brings the title of this book: Lie-Bäcklund-Darboux Transformations which refer to both Bäcklund transformations and Darboux transformations.

The most famous mathematical miracle of the 20th century was probably the discovery of chaos. When the finite dimensional integrable Hamiltonian systems are under perturbations, their regular solutions can turn into chaotic solutions. For such near integrable systems, existence of chaos can sometimes be proved mathematically rigorously. Following the same spirit, one may attempt to prove the existence of chaos for near integrable nonlinear wave equations viewed as near integrable Hamiltonian partial differential equations. This has been accomplished as summarized in the book [69]. The key ingredients in this theory of chaos in partial differential equations are the explicit formulas for the homoclinic orbit and Melnikov integral. The first author's taste is to use Darboux transformation to obtain the homoclinic orbit and Melnikov integral. This will be the focus of the first part of this book.

The second author's taste is to use Darboux transformation in a diversity of applications especially in higher spatial dimensions. The range of applications crosses many different fields of physics. This will be the focus of the second part of this book. This book is a result of the second author's several visits at University of Missouri as a Miller scholar.

The first author would like to thank his wife Sherry and his son Brandon, and the second author would like to thank his wife Alla and his son Valerian, for their loving support during this work.

Contents

Chapter 1

Introduction

The so-called Bäcklund transformation originated from the studies by S. Lie [80–82] and A. V. Bäcklund [11–15] on the Lie's second question on the existence of invariant multi-valued surface transformations [5]. Lie's first question was on the well-known Lie's tangent transformations. The first example of a Bäcklund transformation was studied on the Bianchi's geometrical construction of surfaces of constant negative curvatures – pseudospheres [18]. The Gauss equation of a pseudosphere can be rewritten as the sine-Gordon equation. The Bäcklund transformation for the sine-Gordon equation is an invariant transformation with a so-called Bäcklund parameter first introduced by Bäcklund. The Bäcklund parameter is particularly important in Bianchi's diagram of iterating the Bäcklund transformation to generate a so-called nonlinear superposition law [18, 19]. Immediate further studies on Bäcklund transformations were conducted by J. Clairin [26] and E. Goursat [40].

Darboux transformation was first introduced by Gaston Darboux [29] for the nowadays well-known one-dimensional linear Schrödinger equation – a special form of the Sturm-Liouville equations [84]. Darboux found a covariant transformation for the eigenfunction and the potential. The covariant transformation was built upon a particular eigenfunction at a particular value of the spectral parameter.

At the beginning, it seemed that Bäcklund transformation and Darboux transformation are irrelevant. The first link of the two came about in 1967 when Gardner, Greene, Kruskal, and Miura related KdV equation to its Lax pair of which the spatial part is the one-dimensional linear Schrödinger equation [38]. Soon afterwards, the Bäcklund transformation for the KdV equation was found. This was the beginning of a renaissance of Bäcklund transformations and Darboux transformations. It turned out that the existence of a Lax pair for a nonlinear wave equation, the solvability of the Cauchy problem for the nonlinear wave equation by the inverse scattering transform [38], the existence of a Bäcklund transformation for the nonlinear wave equation, and the existence of a Darboux transformation for the nonlinear wave equation and its Lax pair are closely related (although clear relation is still not known). Up to now, Bäcklund transformations and Darboux transformations for most of the nonlinear wave equations solvable by the inverse scattering transform, have been found [84, 96]. The potential lies at utilizing these transformations. All the earlier books [3, 5, 27, 41, 84, 91, 95, 96] focus on using Bäcklund or Darboux or inverse scattering transformation to construct multi-soliton solutions. Such multi-soliton solutions are defined on the whole spatial space with decaying boundary conditions. When the integrable system is posed with periodic boundary conditions, the solutions are temporally quasi-periodic or periodic or homoclinic. The first part of this book will focus on homoclinic orbit. Chapters 3–9 contain many valuable formulas for homoclinic orbits and Melnikov integrals. Here the Darboux transformations are not only used to generate explicit formulas for the homoclinic orbits but also interlaced with the isospectral theory of the corresponding Lax pairs to generate Melnikov vectors crucial for building the Melnikov integrals. The integrable systems studied in Chapters 3–9 are the so-called canonical systems each of which models a variety of different phenomena. The formulas can be directly used by the readers to study their own near integrable systems. In Chapter 2, we briefly summarize various methods for deriving Bäcklund transformations. These are all "experimental" or "trial-correction" methods. For a more detailed account on these methods, we refer the readers to the book [96]. There are not many methods for deriving Darboux transformations. The commonly used one is the dressing method [84]. Sometimes Chen's method in Chapter 2 can be effective too. Again these are all "trial-correction" methods. Chapters 10–16 contain applications of Darboux transformations in more specific physics problems, and various connections among different systems. Here no specific boundary condition is posed.

The future of Lie-Bäcklund-Darboux transformations is very bright. Besides the potential of their important applications and new transformations, it is possible to broaden their notion and still end up with useful transformations. This broadening process had begun long ago, e.g. the group notion in [5], the jet bundle in [96], and the Moutard transformation in this book. The broadened transformations even reached the Euler equations of fluids [63, 79].

Chapter 2

A Brief Account on Bäcklund Transformations

Bäcklund transformation is an invariant transformation that transforms one solution into another of a differential equation. It usually involves partial derivatives, and is easily solvable once the initial solution is given. Since J. Clairin [26], researchers have been trying to find an effective systematic way of obtaining a Bäcklund transformation. It was not successful. The commonly employed approaches are Clairin's method [5, 26, 96], Chen's method [25, 91], Hirota's bilinear operator approach [96], and Wahlquist-Estabrook procedure [96, 103, 104]. These are all "trail and correction" approaches. The most common use of Bäcklund transformations is to obtain multi-soliton solutions to integrable systems. For a detailed account, we refer the readers to [96]. Here we shall briefly go over the above mentioned derivation methods.

2.1 A Warm-Up Approach

Take the sine-Gordon equation for example

$$u_{xy} = \sin u, \tag{2.1}$$

where u is a real-valued function of two variables x and y. Let us assume that a Bäcklund transformation transforms u into v where v also satisfies the same sine-Gordon equation

$$v_{xy} = \sin v, \tag{2.2}$$

and our goal is to find the Bäcklund transformation. Let $w^+ = \frac{1}{2}(u+v)$ and $w^- = \frac{1}{2}(u-v)$ [59], then w^+ and w^- satisfy

$$w^+_{xy} = \sin w^+ \cos w^-, \quad w^-_{xy} = \cos w^+ \sin w^-. \tag{2.3}$$

We assume that the Bäcklund transformation has the trial form

$$w_x^- = f(w^+), \tag{2.4}$$

then by the second equation of (2.3), one has

$$f' w_y^+ = \cos w^+ \sin w^-. \tag{2.5}$$

By substituting (2.5) into the first equation of (2.3), one has

$$-\frac{w_x^+ \sin w^-}{f'}\left(\frac{f''}{f'}\cos w^+ + \sin w^+\right) + \cos w^- \left(\frac{f}{f'}\cos w^+ - \sin w^+\right) = 0,$$

which can be satisfied if one demands

$$\frac{f''}{f'}\cos w^+ + \sin w^+ = 0, \quad \text{and} \quad \frac{f}{f'}\cos w^+ - \sin w^+ = 0.$$

A solution of this over-determined system can be found

$$f = a \sin w^+,$$

where a is an arbitrary constant called a Bäcklund parameter. (2.4) and (2.5) now take the form

$$w_x^- = a \sin w^+, \quad w_y^+ = \frac{1}{a}\sin w^-. \tag{2.6}$$

(2.6) implies (2.3) which is equivalent to (2.1) and (2.2). (2.6) is the Bäcklund transformation for sine-Gordon equation (2.1). The above approach of derivation was developed in [59]. After experimenting with different forms of assumptions like (2.4), one hopes to find a Bäcklund transformation for a given equation.

2.2 Chen's Method

Chen's method [25] is a quite efficient method. Take the KdV equation for example,

$$u_t + 6uu_x + u_{xxx} = 0,$$

its Lax Pair is given by

$$L\psi = \lambda\psi, \quad \partial_t \psi = A\psi,$$

where

$$L = \partial_x^2 + u, \quad A = -4\partial_x^3 - 3(u\partial_x + \partial_x u).$$

Let $v = \psi_x/\psi$, one gets

$$v_x + v^2 + u = \lambda, \tag{2.7}$$

$$\begin{aligned} -v_t = {} & 4v_{xxx} + 12vv_{xx} + 12v^2 v_x + 12v_x^2 \\ & + 6u_x v + 6uv_x + 3u_{xx}. \end{aligned} \tag{2.8}$$

Eliminating u, one finds that v satisfies the modified KdV equation

$$v_t - 6v^2 v_x + 6\lambda v_x + v_{xxx} = 0.$$

Relation (2.7) is the Miura transformation between KdV and modified KdV. If v solves the modified KdV, so does $-v$, then one can find $\hat{u}$ by solving KdV such that

$$-v_x + v^2 + \hat{u} = \lambda, \tag{2.9}$$

$$\begin{aligned} v_t = &- 4v_{xxx} + 12vv_{xx} - 12v^2 v_x + 12v_x^2 \\ &- 6\hat{u}_x v - 6\hat{u}v_x + 3\hat{u}_{xx}. \end{aligned} \tag{2.10}$$

Subtracting (2.9) from (2.7), one gets $v_x = \frac{1}{2}(\hat{u} - u)$. Let $w_x = \frac{1}{2}u$ and $\hat{w}_x = \frac{1}{2}\hat{u}$, then from (2.7) and (2.8), one gets the Bäcklund transformation for KdV equation

$$\begin{aligned} (\hat{w} + w)_x &= \lambda - (\hat{w} - w)^2, \\ (w - \hat{w})_t &= 4(\hat{w} - w)_{xxx} + 12(\hat{w} - w)(\hat{w} - w)_{xx} \\ &\quad + 12(\hat{w} - w)^2(\hat{w} - w)_x + 12(\hat{w} - w)_x^2 \\ &\quad + 12(\hat{w} - w)w_{xx} + 12w_x(\hat{w} - w)_x + 6w_{xxx}, \end{aligned}$$

where λ is the Bäcklund parameter.

2.3 Clairin's Method

Clairin's method is a direct trial method which involves tedious calculations. In general, let $u = u(t, x)$ satisfy some equations, and let $v = v(t, x)$ be another variable linked to u through a transformation of the form

$$u_x = f(u, v, v_x, v_t), \quad u_t = g(u, v, v_x, v_t).$$

The compatibility condition

$$f_t = g_x,$$

hopefully can lead to an equation for v, of course, by virtue of the equation satisfied by u. Consider the focusing cubic nonlinear Schrödinger equation

$$iq_t + q_{xx} + |q|^2 q = 0,$$

Lamb started with a transformation of the form [55]

$$\begin{aligned} q_x &= f(q, \bar{q}, p, \bar{p}, p_x, \bar{p}_x), \\ q_t &= g(q, \bar{q}, p, \bar{p}, p_x, \bar{p}_x, p_t, \bar{p}_t). \end{aligned}$$

After some lengthy calculation, Lamb obtained the Bäcklund transformation [55],

$$q_x = p_x - \frac{1}{2} i w \xi + i k v,$$
$$q_t = p_t + \frac{1}{2} \xi w_x - k \zeta + \frac{1}{4} i v (|w|^2 + |v|^2),$$

where

$$\xi = \pm i (b - 2|v|^2)^{1/2}, \quad w = q + p,$$
$$v = q - p, \quad \zeta = -\frac{1}{2} i w \xi + i k v,$$

b and k are arbitrary real constants called Bäcklund parameters, and p and q satisfy the same equation.

2.4 Hirota's Bilinear Operator Method

Hirota [45, 46] introduced certain bilinear operators to convert the nonlinear wave equations into bilinear equations for which Bäcklund transformations can be constructed. The Hirota bilinear operators D_t and D_x act according to

$$D_t^m D_x^n f \circ g = (\partial_t - \partial_{\hat{t}})^m (\partial_x - \partial_{\hat{x}})^n f(t,x) g(\hat{t},\hat{x})|_{\hat{t}=t,\hat{x}=x}.$$

Consider the KdV equation

$$u_t + 6 u u_x + u_{xxx} = 0,$$

by setting $u = 2(\ln f)_{xx}$, one obtains

$$u_t + 6 u u_x + u_{xxx} = \partial_x \left[\frac{1}{f^2} D_x (D_t + D_x^3) f \circ f \right].$$

Thus, if f solves the associated bilinear equation

$$D_x (D_t + D_x^3) f \circ f = 0,$$

then u solves the KdV equation. A Bäcklund transformation for the bilinear equation can be obtained as [46]

$$(D_x^2 - \beta) f \circ \tilde{f} = 0, \quad (D_t + 3\beta D_x + D^3) f \circ \tilde{f} = 0,$$

where β is the Bäcklund parameter.

2.5 Wahlquist-Estabrook Procedure

Wahlquist-Estabrook Procedure [103, 104] offers a relatively more systematic approach than that of Clairin [96]. Consider the equation (which is essentially the stationary 2D Euler equation) [58]

$$\Delta u = f(u), \tag{2.11}$$

where $\Delta = \partial_x^2 + \partial_y^2$, and f is an arbitrary function. Let $M = \mathbb{R}^2$ (with coordinates x, y), $N = \mathbb{R}^1$ (with coordinate u), $N' = \mathbb{R}^1$ (with coordinate v), w denote the volume form on M, $w = dx \wedge dy$, and θ denote the contact 1-form on the 1-jet $J^1(M, N)$, $\theta = du - u_x dx - u_y dy$. The exterior differential system of 2-forms on $J^1(M, N)$ associated with (2.11) is generated by

$$\sigma = du_x \wedge dy - du_y \wedge dx - f(u)w,$$
$$\eta_1 = du \wedge dy - u_x dx \wedge dy,$$
$$\eta_2 = du \wedge dx + u_y dx \wedge dy.$$

We seek the following form of Bäcklund transformations for equation (2.11),

$$v_z = \psi_z(u, u_x, u_y, v), \quad (z = x, y).$$

The Wahlquist-Estabrook procedure [96, 103, 104] requires that

$$d\psi_x \wedge dx + d\psi_y \wedge dy = f_1\eta_1 + f_2\eta_2 + g\sigma + \xi \wedge \zeta,$$

where f_1, f_2, and g are arbitrary functions on $J^1(M, N) \times J^0(M, N')$, $\zeta = dv - \psi_x dx - \psi_y dy$, and ξ is a 1-form on $J^1(M, N) \times J^0(M, N')$. Solving this equation, one finds that if f satisfies [58, 98]

$$\frac{d^2 f}{du^2} = \lambda f,$$

for an arbitrary constant λ, then there is a Bäcklund transformation, and u and v satisfy the same equation.

Chapter 3

Nonlinear Schrödinger Equation

In the late 19th century, G. Darboux [29] introduced a type of transformations, now called the Darboux transformations, for Sturm-Liouville systems. The Darboux transformation is a covariant transformation which transforms the solution and the coefficient (potential) simultaneously.

For soliton systems, the corresponding Darboux transformation involves both the evolution equation and its Lax pair. In this context, a Darboux transformation is another form of the Lie-Bäcklund-Darboux transformation. Like the Bäcklund transformations, the derivation method for Darboux transformations is often of "experimental" nature. The popular ones include the "dressing method" [84] and the Chen's method [25]. As shown in the previous chapter, the main application of a Bäcklund transformation is at generating multisoliton solutions. Besides generating multisoliton solutions, a Darboux transformation has another novel application—generating homoclinic (heteroclinic) orbits. This new application was not heavily publicized. Its importance is obvious. Homoclinic (heteroclinic) orbits are the fertile ground of chaos when the system is under perturbations [60–62, 66–72, 75–78]. These homoclinic (heteroclinic) orbits form a figure eight structure also called a separatrix.

We take the focusing nonlinear Schrödinger equation (NLS) as our first example to show how to construct figure eight structures [70, 76]. If one starts from the conservation laws of the NLS, it turns out that it is very difficult to get the separatrices. On the contrary, starting from the Darboux transformation to be presented, one can find the separatrices rather easily.

3.1 Physical Background

The term "nonlinear Schrödinger equation" of course comes from the well-known (linear) Schrödinger equation of quantum mechanics. Folklore says that it was artificially written down at first and then discovered in many physical applications.

There are two types of nonlinear Schrödinger equations: focusing v.s. defocusing.

$$iq_t = q_{xx} \pm 2|q|^2 q$$

where q is a complex-valued function of the two real variables (t, x), the upper '+' sign corresponds to focusing, and the lower '−' sign corresponds to defocusing. More general form may include multi-spatial dimensions with the second derivative replaced by Laplacian, and more general nonlinear terms. In the water wave context, the focusing nonlinear Schrödinger equation models weakly nonlinear surface wave on deep water, while the defocusing nonlinear Schrödinger equation models weakly nonlinear surface wave on shallow water. The focusing nonlinear Schrödinger equation has envelope soliton solutions, and homoclinic solutions under periodic boundary conditions; while the defocusing nonlinear Schrödinger equation has neither. Shallow water solitons are modeled by KdV equation etc. In the fiber optics context, the focusing nonlinear Schrödinger equation models the light wave propagation through a nonlinear medium. An interesting note here is that the t-variable actually represents space, and the x-variable actually represents time. In the ferromagnet context, the focusing nonlinear Schrödinger equation is gauge equivalent to the Heisenberg ferromagnet equation. There are many other contexts where the nonlinear Schrödinger equation appears. In this sense, it is regarded as a canonical equation. The principal novelty of the more general form of the nonlinear Schrödinger equation is the finite-time blow-up solution. It is here for the more general nonlinear Schrödinger equations where the finite-time blow-up solutions are best understood. The nonlinear Schrödinger equation is also closely related to Gross-Pitaevskii equation in Bose-Einstein condensate theory, complex Ginzburg-Landau equation, and Davey-Stewartson equations.

3.2 Lax Pair and Floquet Theory

We consider the NLS

$$iq_t = q_{xx} + 2|q|^2 q, \tag{3.1}$$

under periodic boundary condition $q(x + 2\pi) = q(x)$. The NLS is an integrable system by virtue of the Lax pair [111],

$$\varphi_x = U\varphi, \tag{3.2}$$

$$\varphi_t = V\varphi, \tag{3.3}$$

where

$$U = i\lambda\sigma_3 + i\begin{pmatrix} 0 & q \\ -r & 0 \end{pmatrix},$$

$$V = 2i\lambda^2\sigma_3 + iqr\sigma_3 + \begin{pmatrix} 0 & 2i\lambda q + q_x \\ -2i\lambda r + r_x & 0 \end{pmatrix},$$

where σ_3 denotes the third Pauli matrix $\sigma_3 = \text{diag}(1,-1)$, $r = -\bar{q}$, and λ is the spectral parameter. If q satisfies the NLS, then the compatibility of the over determined system (3.2) and (3.3) is guaranteed. Let $M = M(x)$ be the fundamental matrix solution to the ODE (3.2), $M(0)$ is the 2×2 identity matrix. We introduce the so-called transfer matrix $T = T(\lambda, \vec{q})$ where $\vec{q} = (q, -\bar{q})$, $T = M(2\pi)$.

Lemma 3.1. *Let $Y(x)$ be any solution to the ODE (3.2), then*

$$Y(2n\pi) = T^n \, Y(0).$$

Proof: Since $M(x)$ is the fundamental matrix,

$$Y(x) = M(x) \, Y(0).$$

Thus,

$$Y(2\pi) = T \, Y(0).$$

Assume that

$$Y(2l\pi) = T^l \, Y(0).$$

Notice that $Y(x + 2l\pi)$ also solves the ODE (3.2), then

$$Y(x + 2l\pi) = M(x) \, Y(2l\pi),$$

thus,

$$Y(2(l+1)\pi) = T \, Y(2l\pi) = T^{l+1} \, Y(0).$$

The lemma is proved. □

Definition 3.2. We define the Floquet discriminant Δ as

$$\Delta(\lambda; \vec{q}) = \text{trace}\,(T(\lambda, \vec{q})).$$

We define the periodic and anti-periodic points $\lambda^{(p)}$ by the condition

$$|\Delta(\lambda^{(p)}; \vec{q})| = 2.$$

We define the critical points $\lambda^{(c)}$ by the condition

$$\left.\frac{\partial \Delta(\lambda; \vec{q})}{\partial \lambda}\right|_{\lambda = \lambda^{(c)}} = 0.$$

A multiple point, denoted by $\lambda^{(m)}$, is a critical point for which

$$|\Delta(\lambda^{(m)}; \vec{q})| = 2.$$

The algebraic multiplicity of $\lambda^{(m)}$ is defined as the order of the zero of $\Delta(\lambda) \pm 2$. Usually it is 2, but it can exceed 2; when it does equal 2, we call the multiple point a double point, and denote it by $\lambda^{(d)}$. The geometric multiplicity of $\lambda^{(m)}$ is defined as the maximum number of linearly independent solutions to the ODE (3.2), and is either 1 or 2.

3.3 Darboux Transformations and Homoclinic Orbit

Let $q(x,t)$ be a solution to the NLS (3.1) for which the linear system (3.2) has a complex double point ν of geometric multiplicity 2. We denote two linearly independent solutions of the Lax pair (3.2) and (3.3) at $\lambda = \nu$ by (ϕ^+, ϕ^-). Thus, a general solution of the linear systems at (q, ν) is given by

$$\phi(x,t) = c_+\phi^+ \ + \ c_-\phi^- . \tag{3.4}$$

We use ϕ to define a Gauge matrix G by

$$G = G(\lambda; \nu; \phi) = N \begin{pmatrix} \lambda - \nu & 0 \\ 0 & \lambda - \bar{\nu} \end{pmatrix} N^{-1}, \tag{3.5}$$

where

$$N = \begin{pmatrix} \phi_1 & -\bar{\phi}_2 \\ \phi_2 & \bar{\phi}_1 \end{pmatrix}. \tag{3.6}$$

Then we define Q and Ψ by

$$Q(x,t) = q(x,t) \ + \ 2(\nu - \bar{\nu}) \frac{\phi_1 \bar{\phi}_2}{\phi_1 \bar{\phi}_1 + \phi_2 \bar{\phi}_2} \tag{3.7}$$

and

$$\Psi(x,t;\lambda) \ = \ G(\lambda; \nu; \phi) \ \psi(x,t;\lambda) \tag{3.8}$$

where ψ solves the Lax pair (3.2) and (3.3) at (q, ν). Formulas (3.7) and (3.8) are the Bäcklund-Darboux transformations for the potential and eigenfunctions, respectively. We have the following [76].

Theorem 3.3. *Let $q(x,t)$ be a solution to the NLS equation* (3.1), *for which the linear system* (3.2) *has a complex double point ν of geometric multiplicity* 2, *with eigenbasis (ϕ^+, ϕ^-) for the Lax pair* (3.2) *and* (3.3), *and define $Q(x,t)$ and $\Psi(x,t;\lambda)$ by* (3.7) *and* (3.8). *Then*

(1) *$Q(x,t)$ is a solution of NLS, with spatial period 2π,*

(2) *Q and q have the same Floquet spectrum,*

(3) *$Q(x,t)$ is homoclinic to $q(x,t)$ in the sense that $Q(x,t) \longrightarrow q_{\theta_\pm}(x,t)$, exponentially as $\exp(-\sigma_\nu |t|)$, as $t \longrightarrow \pm\infty$, where $q_{\theta_\pm}$ is a "torus translate" of q, σ_ν is the nonvanishing growth rate associated to the complex double point ν, and explicit formulas exist for this growth rate and for the translation parameters $\theta_\pm$,*

(4) *$\Psi(x,t;\lambda)$ solves the Lax pair* (3.2) *and* (3.3) *at (Q, λ).*

This theorem is quite general, constructing homoclinic solutions from a wide class of starting solutions $q(x,t)$. Its proof is one of direct verification [60].

We emphasize several qualitative features of these homoclinic orbits: (i) $Q(x,t)$ is homoclinic to a torus which itself possesses rather complicated spatial and temporal structure, and is not just a fixed point. (ii) Nevertheless, the homoclinic orbit typically has still more complicated spatial structure than its "target torus". (iii) When there are several complex double points, each with nonvanishing growth rate, one can iterate the Bäcklund-Darboux transformations to generate more complicated homoclinic orbits. (iv) The number of complex double points with nonvanishing growth rates counts the dimension of the unstable manifold of the critical torus in that two unstable directions are coordinatized by the complex ratio c_+/c_-. Under even symmetry only one real dimension satisfies the constraint of evenness, as will be clearly illustrated in the following example. (v) These Bäcklund-Darboux formulas provide global expressions for the stable and unstable manifolds of the critical tori, which represent figure eight structures.

Example: As a concrete example, we take $q(x,t)$ to be the special solution

$$q_c = c \exp\left[-i(2c^2 t + \gamma)\right] . \tag{3.9}$$

Solutions of the Lax pair (3.2) and (3.3) can be computed explicitly:

$$\phi^{(\pm)}(x,t;\lambda) = e^{\pm i\kappa(x+2\lambda t)} \begin{pmatrix} c e^{-i(2c^2t+\gamma)/2} \\ (\pm\kappa - \lambda) e^{i(2c^2t+\gamma)/2} \end{pmatrix}, \tag{3.10}$$

where

$$\kappa = \kappa(\lambda) = \sqrt{c^2 + \lambda^2}.$$

With these solutions one can construct the fundamental matrix

$$M(x;\lambda;q_c) = \begin{pmatrix} \cos\kappa x + i\frac{\lambda}{\kappa}\sin\kappa x & i\frac{q_c}{\kappa}\sin\kappa x \\ i\frac{\overline{q_c}}{\kappa}\sin\kappa x & \cos\kappa x - i\frac{\lambda}{\kappa}\sin\kappa x \end{pmatrix}, \tag{3.11}$$

from which the Floquet discriminant can be computed:

$$\Delta(\lambda;q_c) = 2\cos 2\kappa\pi. \tag{3.12}$$

From Δ, spectral quantities can be computed:
(1) simple periodic points: $\lambda^\pm = \pm i\, c$,
(2) double points: $\kappa(\lambda_j^{(d)}) = j/2,\ j \in \mathbb{Z},\ j \neq 0$,
(3) critical points: $\lambda_j^{(c)} = \lambda_j^{(d)},\ j \in \mathbb{Z},\ j \neq 0$,
(4) simple periodic points: $\lambda_0^{(c)} = 0$.
For this spectral data, there are $2N$ purely imaginary double points,

$$(\lambda_j^{(d)})^2 = j^2/4 - c^2, \quad j = 1, 2, \ldots, N; \tag{3.13}$$

where

$$(N^2/4 - c^2) < 0 < ((N+1)^2/4 - c^2).$$

From this spectral data, the homoclinic orbits can be explicitly computed through Bäcklund-Darboux transformation. Notice that to have temporal growth (and decay) in the eigenfunctions (3.10), one needs λ to be complex. Notice also that the Bäcklund-Darboux transformation is built with quadratic products in ϕ, thus choosing $\nu = \lambda_j^{(d)}$ will guarantee periodicity of Q in x. When $N = 1$, the Bäcklund-Darboux transformation at one purely imaginary double point $\lambda_1^{(d)}$ yields $Q = Q(x, t; c, \gamma; c_+/c_-)$ [76]:

$$\begin{aligned} Q &= (\cos 2p - \sin p \ \mathrm{sech}\ \tau \ \cos(x+\vartheta) - i \sin 2p \tanh \tau) \\ &\quad \cdot (1 + \sin p \ \mathrm{sech}\ \tau \ \cos(x+\vartheta))^{-1} c e^{-i(2c^2 t + \gamma)} \\ &\to e^{\mp 2ip} c e^{-i(2c^2 t+\gamma)} \qquad \text{as } \ \rho \to \mp\infty, \end{aligned} \tag{3.14}$$

where $c_+/c_- \equiv \exp(\rho + i\beta)$ and p is defined by $1/2 + i\sqrt{c^2 - 1/4} = c \exp(ip)$, $\tau \equiv \sigma t - \rho$, and $\vartheta = p - (\beta + \pi/2)$.

Several points about this homoclinic orbit need to be made:

(1) The orbit depends only upon the ratio c_+/c_-, and not upon c_+ and c_- individually.

(2) Q is homoclinic to the plane wave orbit; however, a phase shift of $-4p$ occurs when one compares the asymptotic behavior of the orbit as $t \to -\infty$ with its behavior as $t \to +\infty$.

(3) For small p, the formula for Q becomes more transparent:

$$Q \simeq ((\cos 2p - i \sin 2p \tanh \tau) - 2 \sin p \ \mathrm{sech}\ \tau \cos(x+\vartheta)) c e^{-i(2c^2 t+\gamma)}.$$

(4) An evenness constraint on Q in x can be enforced by restricting the phase ϕ to be one of two values

$$\phi = 0, \pi. \quad \text{(evenness)}$$

In this manner, the even symmetry disconnects the level set. Each component constitutes one loop of the figure eight. While the target q is independent of x, each of these loops has x dependence through the $\cos x$. One loop has exactly this dependence and can be interpreted as a spatial excitation located near $x = 0$, while the second loop has the dependence $\cos(x - \pi)$, which we interpret as spatial structure located near $x = \pi$. In this example, the disconnected nature of the level set is clearly related to distinct spatial structures on the individual loops. See Figure 1 for an illustration.

(5) Direct calculation shows that the transformation matrix $M(1; \lambda_1^{(d)}; Q)$ is similar to a Jordan form when $t \in (-\infty, \infty)$,

$$M(1; \lambda_1^{(d)}; Q) \sim \begin{pmatrix} -1 & 1 \\ 0 & -1 \end{pmatrix},$$

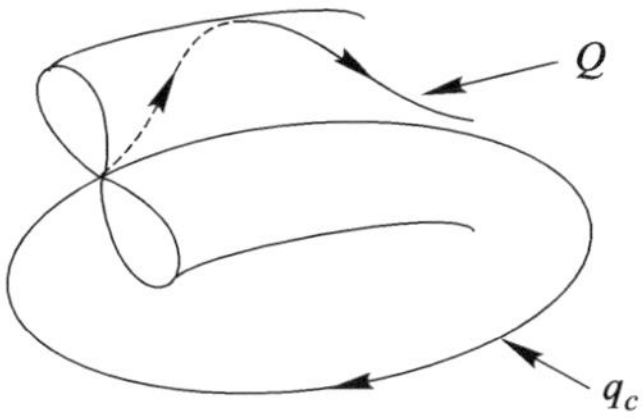

Figure 1 An illustration of the figure-eight structure.

and when $t \to \pm\infty$, $M(1;\lambda_1^{(d)};Q) \longrightarrow -I$ (the negative of the 2×2 identity matrix). Thus, when t is finite, the algebraic multiplicity ($=2$) of $\lambda = \lambda_1^{(d)}$ with the potential Q is greater than the geometric multiplicity ($=1$).

In this example the dimension of the loops needs not be one, but is determined by the number of purely imaginary double points which in turn is controlled by the amplitude c of the plane wave target and by the spatial period. (The dimension of the loops increases linearly with the spatial period.) When there are several complex double points, Bäcklund-Darboux transformations must be iterated to produce complete representations. Thus, Bäcklund-Darboux transformations give global representations of the figure-eight structures.

3.4 Linear Instability

The above figure-eight structure corresponds to the following linear instability of Benjamin-Feir type. Consider the uniform solution to the NLS (3.1),

$$q_c = ce^{i\theta(t)}, \quad \theta(t) = -(2c^2 t + \gamma).$$

Let

$$q = (c + \tilde{q})e^{i\theta(t)},$$

and linearize equation (3.1) at q_c, we have

$$i\tilde{q}_t = \tilde{q}_{xx} + 2c^2(\tilde{q} + \bar{\tilde{q}}).$$

Assume that $\tilde{q}$ takes the form

$$\tilde{q} = (A_j e^{\Omega_j t} + B_j e^{\bar{\Omega}_j t})\cos(k_j x),$$

where $k_j = 2j\pi$ ($j = 0, 1, 2, \ldots$), A_j and B_j are complex constants. Then,

$$\Omega_j^{(\pm)} = \pm k_j\sqrt{4c^2 - k_j^2}.$$

Thus, we have instabilities when $c > 1/2$.

3.5 Quadratic Products of Eigenfunctions

Quadratic products of eigenfunctions play a crucial role in characterizing the hyperbolic structures of soliton equations. Its importance lies in the following aspects: (i) Certain quadratic products of eigenfunctions solve the linearized soliton equation. (ii) Thus, they are the perfect candidates for building a basis to the invariant linear subbundles. (iii) Also, they signify the instability of the soliton equation. (iv) Most importantly, quadratic products of eigenfunctions can serve as Melnikov vectors, e.g., for Davey-Stewartson equation [62].

Consider the linearized NLS equation at any solution $q(t,x)$ written in the vector form:

$$\begin{aligned} i\partial_t(\delta q) &= (\delta q)_{xx} + 2(q^2\overline{\delta q} + 2|q|^2\delta q), \\ i\partial_t(\overline{\delta q}) &= -(\overline{\delta q})_{xx} - 2(\bar{q}^2\delta q + 2|q|^2\overline{\delta q}), \end{aligned} \tag{3.15}$$

we have the following lemma [76].

Lemma 3.4. *Let* $\varphi^{(j)} = \varphi^{(j)}(t,x;\lambda,q)$ $(j=1,2)$ *be any two eigenfunctions solving the Lax pair* (3.2) *and* (3.3) *at an arbitrary* λ. *Then*

$$\begin{pmatrix} \delta q \\ \overline{\delta q} \end{pmatrix}, \quad \begin{pmatrix} \varphi_1^{(1)} & \varphi_1^{(2)} \\ \varphi_2^{(1)} & \varphi_2^{(2)} \end{pmatrix}, \quad \text{and } S \begin{pmatrix} \varphi_1^{(1)} & \varphi_1^{(2)} \\ \varphi_2^{(1)} & \varphi_2^{(2)} \end{pmatrix}^-, \quad \text{where } S = \begin{pmatrix} 0 & 1 \\ 1 & 0 \end{pmatrix}$$

solve the same equation (3.15)*; thus*

$$\Phi = \begin{pmatrix} \varphi_1^{(1)} & \varphi_1^{(2)} \\ \varphi_2^{(1)} & \varphi_2^{(2)} \end{pmatrix} + S \begin{pmatrix} \varphi_1^{(1)} & \varphi_1^{(2)} \\ \varphi_2^{(1)} & \varphi_2^{(2)} \end{pmatrix}^-$$

solves equation (3.15) *and satisfies the reality condition* $\Phi_2 = \bar{\Phi}_1$.

Proof: Direct calculation leads to the conclusion. □

The periodicity condition $\Phi(x+2\pi) = \Phi(x)$ can be easily accomplished. For example, we can take $\varphi^{(j)}$ $(j=1,2)$ to be two linearly independent Bloch functions $\varphi^{(j)} = e^{\sigma_j x}\psi^{(j)}$ $(j=1,2)$, where $\sigma_2 = -\sigma_1$ and $\psi^{(j)}$ are periodic functions $\psi^{(j)}(x+2\pi) = \psi^{(j)}(x)$. Often we choose λ to be a double point of geometric multiplicity 2, so that $\varphi^{(j)}$ are already periodic or antiperiodic functions.

3.6 Melnikov Vectors

Definition 3.5. Define the sequence of functionals F_j as follows,

$$F_j(\vec{q}) = \Delta(\lambda_j^c(\vec{q}); \vec{q}), \tag{3.16}$$

where λ_j^c's are the critical points, $\vec{q} = (q, -\bar{q})$.

We have the lemma [76]:

Lemma 3.6. *If $\lambda_j^c(\vec{q})$ is a simple critical point of $\Delta(\lambda)$ [i.e., $\Delta''(\lambda_j^c) \neq 0$], F_j is analytic in a neighborhood of $\vec{q}$, with first derivative given by*

$$\frac{\delta F_j}{\delta q} = \frac{\delta \Delta}{\delta q}\bigg|_{\lambda=\lambda_j^c} + \frac{\partial \Delta}{\partial \lambda}\bigg|_{\lambda=\lambda_j^c} \frac{\delta \lambda_j^c}{\delta q} = \frac{\delta \Delta}{\delta q}\bigg|_{\lambda=\lambda_j^c}, \tag{3.17}$$

where

$$\frac{\delta}{\delta \vec{q}}\, \Delta(\lambda; \vec{q}) = i \frac{\sqrt{\Delta^2 - 4}}{W(\psi^+, \psi^-)} \begin{pmatrix} \psi_2^+(x;\lambda)\psi_2^-(x;\lambda) \\ \psi_1^+(x;\lambda)\psi_1^-(x;\lambda) \end{pmatrix}, \tag{3.18}$$

and the Bloch eigenfunctions $\psi^{\pm}$ have the property that

$$\psi^{\pm}(x + 2\pi; \lambda) = \rho^{\pm 1} \psi^{\pm}(x; \lambda), \tag{3.19}$$

for some ρ, also the Wronskian is given by

$$W(\psi^+, \psi^-) = \psi_1^+ \psi_2^- - \psi_2^+ \psi_1^- .$$

In addition, Δ' is given by

$$\frac{d\Delta}{d\lambda} = -i \frac{\sqrt{\Delta^2 - 4}}{W(\psi^+, \psi^-)} \int_0^{2\pi} (\psi_1^+ \psi_2^- + \psi_2^+ \psi_1^-)\, dx. \tag{3.20}$$

Proof: To prove this lemma, one calculates using variation of parameters:

$$\delta M(x; \lambda) = M(x) \int_0^x M^{-1}(x') \delta \hat{Q}(x') M(x') dx',$$

where

$$\delta \hat{Q} \equiv i \begin{pmatrix} 0 & \delta q \\ \delta \bar{q} & 0 \end{pmatrix}.$$

Thus, one obtains the formula

$$\delta \Delta(\lambda; \vec{q}) = \text{ trace } \left[M(2\pi) \int_0^{2\pi} M^{-1}(x')\, \delta \hat{Q}(x')\, M(x') dx' \right],$$

which gives

$$\begin{aligned} \frac{\delta \Delta(\lambda)}{\delta q(x)} &= i \text{ trace } \left[M^{-1}(x) \begin{pmatrix} 0 & 1 \\ 0 & 0 \end{pmatrix} M(x) M(2\pi) \right], \\ \frac{\delta \Delta(\lambda)}{\delta \bar{q}(x)} &= i \text{ trace } \left[M^{-1}(x) \begin{pmatrix} 0 & 0 \\ 1 & 0 \end{pmatrix} M(x) M(2\pi) \right]. \end{aligned} \tag{3.21}$$

Next, we use the Bloch eigenfunctions $\{\psi^{\pm}\}$ to form the matrix

$$N(x;\lambda) = \begin{pmatrix} \psi_1^+ & \psi_1^- \\ \psi_2^+ & \psi_2^- \end{pmatrix}.$$

Clearly,

$$N(x;\lambda) = M(x;\lambda)N(0;\lambda) ;$$

or equivalently,

$$M(x;\lambda) = N(x;\lambda)[N(0;\lambda)]^{-1}. \tag{3.22}$$

Since $\psi^{\pm}$ are Bloch eigenfunctions, one also has

$$N(x+2\pi;\lambda) = N(x;\lambda)\begin{pmatrix} \rho & 0 \\ 0 & \rho^{-1} \end{pmatrix},$$

which implies

$$N(2\pi;\lambda) = M(2\pi;\lambda)N(0;\lambda) = N(0;\lambda)\begin{pmatrix} \rho & 0 \\ 0 & \rho^{-1} \end{pmatrix},$$

that is,

$$M(2\pi;\lambda) = N(0;\lambda)\begin{pmatrix} \rho & 0 \\ 0 & \rho^{-1} \end{pmatrix}[N(0;\lambda)]^{-1}. \tag{3.23}$$

For any 2×2 matrix σ, equations (3.22) and (3.23) imply

$$\text{trace}\,\{[M(x)]^{-1}\ \sigma\ M(x)\ M(2\pi)\} = \text{trace}\,\{[N(x)]^{-1}\ \sigma\ N(x)\ \text{diag}(\rho,\rho^{-1})\},$$

which, through an explicit evaluation of (3.21), proves formula (3.18). Formula (3.20) is established similarly. These formulas, together with the fact that $\lambda^c(\vec{q})$ is differentiable because it is a simple zero of Δ', provide the representation of $\frac{\delta F_j}{\delta \vec{q}}$. □

Remark 3.7. Formula (3.17) for the $\frac{\delta F_j}{\delta \vec{q}}$ is actually valid even if $\Delta''(\lambda_j^c(\vec{q});\vec{q}) = 0$. Consider a function $\vec{q}_*$ at which

$$\Delta''(\lambda_j^c(\vec{q}_*);\vec{q}_*) = 0,$$

and thus at which $\lambda_j^c(\vec{q})$ fails to be analytic. For $\vec{q}$ near $\vec{q}_*$, one has

$$\Delta'(\lambda_j^c(\vec{q});\vec{q}) = 0;$$

$$\frac{\delta}{\delta \vec{q}}\Delta'(\lambda_j^c(\vec{q});\vec{q}) = \Delta''\frac{\delta}{\delta \vec{q}}\lambda_j^c + \frac{\delta}{\delta \vec{q}}\Delta' = 0;$$

that is,

$$\frac{\delta}{\delta \vec{q}}\lambda_j^c = -\frac{1}{\Delta''}\frac{\delta}{\delta \vec{q}}\Delta'.$$

Thus,

$$\frac{\delta}{\delta \vec{q}} F_j = \frac{\delta}{\delta \vec{q}} \Delta + \Delta' \frac{\delta}{\delta \vec{q}} \lambda_j^c = \frac{\delta}{\delta \vec{q}} \Delta - \frac{\Delta'}{\Delta''} \frac{\delta}{\delta \vec{q}} \Delta' |_{\lambda = \lambda_j^c(\vec{q})} .$$

Since $\frac{\Delta'}{\Delta''} \to 0$, as $\vec{q} \to \vec{q}_*$, one still has formula (3.17) at $\vec{q} = \vec{q}_*$:

$$\frac{\delta}{\delta \vec{q}} F_j = \frac{\delta}{\delta \vec{q}} \Delta |_{\lambda = \lambda_j^c(\vec{q})} .$$

The NLS (3.1) is a Hamiltonian system:

$$i q_t = \frac{\delta H}{\delta \bar{q}}, \tag{3.24}$$

where

$$H = \int_0^{2\pi} (-|q_x|^2 + |q|^4) \, dx.$$

Corollary 3.8. *For any fixed* $\lambda \in \mathbb{C}$, $\Delta(\lambda; \vec{q})$ *is a constant of motion of the NLS* (3.1). *In fact,*

$$\{\Delta(\lambda; \vec{q}), H(\vec{q})\} = 0, \quad \{\Delta(\lambda; \vec{q}), \Delta(\lambda'; \vec{q})\} = 0, \quad \forall \lambda, \lambda' \in \mathbb{C},$$

where for any two functionals E *and* F, *their Poisson bracket is defined as*

$$\{E, F\} = \int_0^{2\pi} \left(\frac{\delta E}{\delta q} \frac{\delta F}{\delta \bar{q}} - \frac{\delta E}{\delta \bar{q}} \frac{\delta F}{\delta q} \right) dx.$$

Proof: The corollary follows from a direction calculation from the spatial part (3.2) of the Lax pair and the representation (3.18). □

For each fixed $\vec{q}$, Δ is an entire function of λ; therefore, can be determined by its values at a countable number of values of λ. The invariance of Δ characterizes the isospectral nature of the NLS equation.

Corollary 3.9. *The functionals* F_j *are constants of motion of the NLS* (3.1). *Their gradients provide Melnikov vectors:*

$$\text{grad } F_j(\vec{q}) = i \frac{\sqrt{\Delta^2 - 4}}{W(\psi^+, \psi^-)} \begin{pmatrix} \psi_2^+(x; \lambda_j^c) \psi_2^-(x; \lambda_j^c) \\ \psi_1^+(x; \lambda_j^c) \psi_1^-(x; \lambda_j^c) \end{pmatrix} . \tag{3.25}$$

The distribution of the critical points λ_j^c are described by the following counting lemma [76].

Lemma 3.10 (Counting Lemma for Critical Points). *For* $q \in H^1$, *set* $N = N(\|q\|_1) \in \mathbb{Z}^+$ *by*

$$N(\|q\|_1) = 2[\|q\|_0^2 \cosh(\|q\|_0) \; + \; 3\|q\|_1 \sinh(\|q\|_0)],$$

where $[x]$ = *the first integer greater than* x. *Consider*

$$\Delta'(\lambda;\vec{q}) = \frac{d}{d\lambda}\,\Delta(\lambda;\vec{q}).$$

Then

(1) $\Delta'(\lambda;\vec{q})$ *has exactly* $2N+1$ *zeros (counted according to multiplicity) in the interior of the disc* $D = \{\lambda \in \mathbb{C}: \ |\lambda| < (2N+1)\frac{\pi}{2}\}$;

(2) $\forall k \in \mathbb{Z}, |k| > N, \Delta'(\lambda,\vec{q})$ *has exactly one zero in each disc*

$$\left\{\lambda \in \mathbb{C}: \ |\lambda - k\pi| < \frac{\pi}{4}\right\}.$$

(3) $\Delta'(\lambda;\vec{q})$ *has no other zeros.*

(4) *For* $|\lambda| > (2N+1)\frac{\pi}{2}$, *the zeros of* $\Delta', \{\lambda_j^c, |j| > N\}$, *are all real, simple, and satisfy the asymptotics*

$$\lambda_j^c = j\pi + o(1), \quad \textit{as } |j| \to \infty.$$

3.7 Melnikov Integrals

When studying perturbed integrable systems, the figure-eight structures often lead to chaotic dynamics through homoclinic bifurcations. An extremely powerful tool for detecting homoclinic orbits is the so-called Melnikov integral method [85], which uses "Melnikov integrals" to provide estimates of the distance between the center-unstable manifold and the center-stable manifold of a normally hyperbolic invariant manifold. The Melnikov integrals are often integrals in time of the inner products of certain Melnikov vectors with the perturbations in the perturbed integrable systems. This implies that the Melnikov vectors play a key role in the Melnikov integral method. First, we consider the case of one unstable mode associated with a complex double point ν, for which the homoclinic orbit is given by Bäcklund-Darboux formula (3.7),

$$Q(x,t) \equiv q(x,t) \ + \ 2(\nu - \bar{\nu})\ \frac{\phi_1\bar{\phi}_2}{\phi_1\bar{\phi}_1 + \phi_2\bar{\phi}_2},$$

where q lies in a normally hyperbolic invariant manifold and ϕ denotes a general solution to the Lax pair (3.2) and (3.3) at (q,ν), $\phi = c_+\phi^+ + c_-\phi^-$, and $\phi^\pm$ are Bloch eigenfunctions. Next, we consider the perturbed NLS,

$$iq_t = q_{xx} + 2|q|^2 q + i\epsilon f,$$

where ϵ is the perturbation parameter and f may depend on q and $\bar{q}$ and their spatial derivatives. The *Melnikov integral* can be defined using the constant of motion F_j, where $\lambda_j^c = \nu$ [76]:

$$M_j \equiv \int_{-\infty}^{+\infty}\int_0^{2\pi}\left(\frac{\delta F_j}{\delta q} f + \frac{\delta F_j}{\delta \bar{q}}\bar{f}\right)_{q=Q} dx dt, \tag{3.26}$$

where the integrand is evaluated along the unperturbed homoclinic orbit $q = Q$, and the Melnikov vector $\frac{\delta F_j}{\delta \vec{q}}$ has been given in the last section, which can be expressed rather explicitly using the Bäcklund-Darboux transformation [76] . We begin with the expression (3.25)

$$\frac{\delta F_j}{\delta \vec{q}} = i \frac{\sqrt{\Delta^2 - 4}}{W(\Phi^+, \Phi^-)} \begin{pmatrix} \Phi_2^+ \Phi_2^- \\ \Phi_1^+ \Phi_1^- \end{pmatrix}, \tag{3.27}$$

where $\Phi^{\pm}$ are Bloch eigenfunctions at (Q, ν), which can be obtained from Bäcklund-Darboux formula (3.7):

$$\Phi^{\pm}(x, t; \nu) \equiv G(\nu; \nu; \phi)\ \phi^{\pm}(x, t; \nu),$$

with the transformation matrix G given by

$$G = G(\lambda; \nu; \phi) = N \begin{pmatrix} \lambda - \nu & 0 \\ 0 & \lambda - \bar{\nu} \end{pmatrix} N^{-1},$$

$$N \equiv \begin{pmatrix} \phi_1 & -\bar{\phi}_2 \\ \phi_2 & \bar{\phi}_1 \end{pmatrix}.$$

These Bäcklund-Darboux formulas are rather easy to manipulate to obtain explicit information. For example, the transformation matrix $G(\lambda, \nu)$ has a simple limit as $\lambda \to \nu$:

$$\lim_{\lambda \to \nu} G(\lambda; \nu) = \frac{\nu - \bar{\nu}}{|\phi|^2} \begin{pmatrix} \phi_2 \bar{\phi}_2 & -\phi_1 \bar{\phi}_2 \\ -\phi_2 \bar{\phi}_1 & \phi_1 \bar{\phi}_1 \end{pmatrix}, \tag{3.28}$$

where $|\phi|^2$ is defined by

$$|\vec{\phi}|^2 \equiv \phi_1 \bar{\phi}_1 + \phi_2 \bar{\phi}_2.$$

With formula (3.28), one quickly calculates

$$\Phi^{\pm} = \pm c_{\mp}\ W(\phi^+, \phi^-)\ \frac{\nu - \bar{\nu}}{|\phi|^2} \begin{pmatrix} \bar{\phi}_2 \\ -\bar{\phi}_1 \end{pmatrix},$$

from which one sees that Φ^+ and Φ^- are linearly dependent at (Q, ν),

$$\Phi^+ = -\frac{c_-}{c_+}\ \Phi^-.$$

Remark 3.11. For Q on the figure-eight, the two Bloch eigenfunctions $\Phi^{\pm}$ are linearly dependent. Thus, the geometric multiplicity of ν is only one, even though its algebraic multiplicity is two or higher.

Using L'Hospital's rule, one gets

$$\frac{\sqrt{\Delta^2 - 4}}{W(\Phi^+, \Phi^-)} = \frac{\sqrt{\Delta(\nu)\Delta''(\nu)}}{(\nu - \bar{\nu})\ W(\phi^+, \phi^-)}. \tag{3.29}$$

With formulas (3.27)–(3.29), one obtains the explicit representation of the grad F_j [76]:

$$\frac{\delta F_j}{\delta \vec{q}} = C_\nu \frac{c_+ c_- W(\psi^{(+)}, \psi^{(-)})}{|\phi|^4} \begin{pmatrix} \bar{\phi}_2^2 \\ -\bar{\phi}_1^2 \end{pmatrix}, \tag{3.30}$$

where the constant C_ν is given by

$$C_\nu \equiv i(\nu - \bar{\nu}) \ \sqrt{\Delta(\nu)\Delta''(\nu)}.$$

With these ingredients, one obtains the following beautiful representation of the Melnikov function associated to the general complex double point ν [76]:

$$M_j = C_\nu \ c_+ c_- \int_{-\infty}^{+\infty} \int_0^{2\pi} \ W(\phi^+, \phi^-) \ \left[\frac{(\bar{\phi}_2^2) f(Q, \bar{Q}) + (\bar{\phi}_1^2) \overline{f(Q, \bar{Q})}}{|\phi|^4} \right] dx dt. \tag{3.31}$$

In the case of several complex double points, each associated with an instability, one can iterate the Bäcklund-Darboux transformations and use those functionals F_j which are associated with each complex double point to obtain representations *Melnikov Vectors.* In general, the relation between $\frac{\delta F_j}{\delta \vec{q}}$ and double points can be summarized in the following lemma [76].

Lemma 3.12. *Except for the trivial case* $q = 0$,
(1) $\frac{\delta F_j}{\delta q} = 0 \Leftrightarrow \frac{\delta F_j}{\delta \bar{q}} = 0 \Leftrightarrow M(2\pi, \lambda_j^c; \ \vec{q}_*) = \pm I.$
(2) $\left.\frac{\delta F_j}{\delta \vec{q}}\right|_{\vec{q}_*} = 0 \Rightarrow \Delta'(\lambda_j^c(\vec{q}_*); \ \vec{q}_*) = 0, \Rightarrow |F_j(\vec{q}_*)| = 2,$
$\Rightarrow \lambda_j^c(\vec{q}_*)$ *is a multiple point,*
where I *is the* 2×2 *identity matrix.*

The Bäcklund-Darboux transformation theorem indicates that the figure-eight structure is attached to a complex double point. The above lemma shows that at the origin of the figure-eight, the gradient of F_j vanishes. Together they indicate that the the gradient of F_j along the figure-eight is a perfect Melnikov vector.

Example: When $\frac{1}{2} < c < 1$ in (3.9), and choosing $\vartheta = \pi$ in (3.14), one can get the Melnikov vector field along the homoclinic orbit (3.14),

$$\begin{aligned} \frac{\delta F_1}{\delta q} &= 2\pi \ \sin^2 p \ \mathrm{sech} \ \tau \ \frac{[(-\sin p + i \cos p \tanh \tau) \cos x + \ \mathrm{sech} \ \tau]}{(1 - \sin p \ \mathrm{sech} \ \tau \cos x)^2} \ c \, e^{i\theta}, \\ \frac{\delta F_1}{\delta \bar{q}} &= \overline{\frac{\delta F_1}{\delta q}}. \end{aligned} \tag{3.32}$$

Chapter 4

Sine-Gordon Equation

4.1 Background

Sine-Gordon equation was the first equation for which a Bäcklund transformation was found. The first application of the sine-Gordon equation was in differential geometry. It turned out that the Gauss equation of a pseudosphere can be rewritten as the sine-Gordon equation [18]. It was in the pseudosphere study, the first Bäcklund transformation was discovered and the first Bianchi diagram was developed to build a nonlinear superposition principle for the solutions of the sine-Gordon equation [15, 18]. For a more detailed account, we refer the readers to [5]. Modern applications of the sine-Gordon equation and its various generalizations have an extremely wide scope ranging from physics to biology. For instance, the flux dynamics in a long Josephson junction is described by the sine-Gordon equation. The "sine-Gordon" was coined from the well-known Klein-Gordon equation. It is closely connected to other integrable systems like elliptic sine-Gordon equation, sinh-Gordon equation, and elliptic sinh-Gordon equation.

4.2 Lax Pair

Here we consider the sine-Gordon equation

$$u_{tt} = c^2 u_{xx} + \sin u, \quad (c \text{ is a positive constant}) \tag{4.1}$$

which is integrable through the Lax pair

$$\psi_x = B\psi, \tag{4.2}$$

$$\psi_t = A\psi, \tag{4.3}$$

where

$$B = \frac{1}{c} \begin{pmatrix} \frac{i}{4}(cu_x + u_t) & \frac{1}{16\lambda} e^{iu} + \lambda \\ -\frac{1}{16\lambda} e^{-iu} - \lambda & -\frac{i}{4}(cu_x + u_t) \end{pmatrix},$$

$$A = \begin{pmatrix} \frac{i}{4}(cu_x + u_t) & -\frac{1}{16\lambda} e^{iu} + \lambda \\ \frac{1}{16\lambda} e^{-iu} - \lambda & -\frac{i}{4}(cu_x + u_t) \end{pmatrix}.$$

The Lax pair (4.2) and (4.3) possess a symmetry.

Lemma 4.1. *If $\psi = (\psi_1, \psi_2)^T$ solves the Lax pair (4.2) and (4.3) at (λ, u), then $(\overline{\psi_2}, \overline{\psi_1})^T$ solves the Lax pair (4.2) and (4.3) at $(-\bar{\lambda}, u)$.*

For more details on this chapter, see [68, 73].

4.3 Darboux Transformations

There is a Darboux transformation for the Lax pair (4.2) and (4.3).

Theorem 4.2 (Darboux Transformation I). *Let*

$$U = u + 2i \ln \left(\frac{i\phi_2}{\phi_1} \right),$$

$$\Psi = \begin{pmatrix} -\nu\phi_2/\phi_1 & \lambda \\ -\lambda & \nu\phi_1/\phi_2 \end{pmatrix} \psi,$$

where $\phi = \psi|_{\lambda=\nu}$ for some ν, then Ψ solves the Lax pair (4.2) and (4.3) at (λ, U).

Often in order to guarantee the reality condition (i.e., U needs to be real-valued), one needs to iterate the Darboux transformation by virtue of Lemma 4.1. The result corresponds to the counterpart of the Darboux transformation for the cubic nonlinear Schrödinger equation [68, 76].

Theorem 4.3 (Darboux Transformation II). *Let*

$$U = u + 2i \ln \left(\frac{\nu|\phi_1|^2 + \bar{\nu}|\phi_2|^2}{\bar{\nu}|\phi_1|^2 + \nu|\phi_2|^2} \right),$$

$$\Psi = G\psi,$$

where

$$G = \begin{pmatrix} G_1 & G_2 \\ G_3 & G_4 \end{pmatrix},$$

$$G_1 = |\nu|^2 \frac{\nu|\phi_1|^2 + \bar{\nu}|\phi_2|^2}{\bar{\nu}|\phi_1|^2 + \nu|\phi_2|^2} - \lambda^2,$$

$$G_2 = \frac{\lambda(\nu^2 - \bar{\nu}^2)\phi_1\overline{\phi_2}}{\bar{\nu}|\phi_1|^2 + \nu|\phi_2|^2},$$

$$G_3 = \frac{\lambda(\nu^2 - \bar{\nu}^2)\overline{\phi_1}\phi_2}{\nu|\phi_1|^2 + \bar{\nu}|\phi_2|^2},$$

$$G_4 = |\nu|^2 \frac{\bar{\nu}|\phi_1|^2 + \nu|\phi_2|^2}{\nu|\phi_1|^2 + \bar{\nu}|\phi_2|^2} - \lambda^2,$$

and $\phi = \psi|_{\lambda=\nu}$ *for some* ν, *then* Ψ *solves the Lax pair* (4.2) *and* (4.3) *at* (λ, U).

These Darboux transformations will be used to generate separatrices (figure-eight structures).

4.4 Melnikov Vectors

Focusing upon the spatial part (4.2) of the Lax pair, one can develop a complete Floquet theory. Let $M(x)$ be the fundamental matrix of (4.2), $M(0) = I$ (2×2 identity matrix), then the Floquet discriminant is given as

$$\Delta = \text{trace } M(2\pi).$$

The Floquet spectrum is given by

$$\sigma = \{\lambda \in \mathbb{C} | -2 \le \Delta(\lambda) \le 2\}.$$

Periodic and anti-periodic points $\lambda^\pm$ (which correspond to periodic and anti-periodic eigenfunctions respectively) are defined by

$$\Delta(\lambda^\pm) = \pm 2.$$

A critical point $\lambda^{(c)}$ is defined by

$$\frac{d\Delta}{d\lambda}(\lambda^{(c)}) = 0.$$

A multiple point $\lambda^{(m)}$ is a periodic or anti-periodic point which is also a critical point. The algebraic multiplicity of $\lambda^{(m)}$ is defined as the order of the zero of $\Delta(\lambda) \pm 2$ at $\lambda^{(m)}$. When the order is 2, we call the multiple point a double point, and denote it by $\lambda^{(d)}$. The order can exceed 2. The geometric multiplicity of $\lambda^{(m)}$ is defined as the dimension of the periodic or anti-periodic eigenspace at $\lambda^{(m)}$, and is either 1 or 2.

Counting lemmas for $\lambda^\pm$ and $\lambda^{(c)}$ can be established as in [76, 90], which lead to the existence of the sequences $\{\lambda_j^\pm\}$ and $\{\lambda_j^{(c)}\}$ and their approximate locations. Nevertheless, counting lemmas are not necessary here. For any $\lambda \in \mathbb{C}$, $\Delta(\lambda)$ is invariant under the sine-Gordon flow. This is the so-called isospectral theory.

Definition 4.4. An important sequence of invariants F_j of the sine-Gordon equation is defined by

$$F_j(u, u_t) = \Delta(\lambda_j^{(c)}(u, u_t), u, u_t).$$

Lemma 4.5. *If* $\{\lambda_j^{(c)}\}$ *is a simple critical point of* Δ, *then*

$$\frac{\partial F_j}{\partial w} = \left.\frac{\partial \Delta}{\partial w}\right|_{\lambda=\lambda_j^{(c)}}, \quad w = u, u_t.$$

Proof: We know that

$$\frac{\partial F_j}{\partial w} = \left.\frac{\partial \Delta}{\partial w}\right|_{\lambda=\lambda_j^{(c)}} + \left.\frac{\partial \Delta}{\partial \lambda}\right|_{\lambda=\lambda_j^{(c)}} \frac{\partial \lambda_j^{(c)}}{\partial w}.$$

Since

$$\left.\frac{\partial \Delta}{\partial \lambda}\right|_{\lambda=\lambda_j^{(c)}} = 0,$$

we have

$$\left.\frac{\partial^2 \Delta}{\partial \lambda^2}\right|_{\lambda=\lambda_j^{(c)}} \frac{\partial \lambda_j^{(c)}}{\partial w} + \left.\frac{\partial^2 \Delta}{\partial \lambda \partial w}\right|_{\lambda=\lambda_j^{(c)}} = 0.$$

Since $\lambda_j^{(c)}$ is a simple critical point of Δ,

$$\left.\frac{\partial^2 \Delta}{\partial \lambda^2}\right|_{\lambda=\lambda_j^{(c)}} \neq 0.$$

Thus

$$\frac{\partial \lambda_j^{(c)}}{\partial w} = -\left(\left.\frac{\partial^2 \Delta}{\partial \lambda^2}\right|_{\lambda=\lambda_j^{(c)}}\right)^{-1} \left.\frac{\partial^2 \Delta}{\partial \lambda \partial w}\right|_{\lambda=\lambda_j^{(c)}}.$$

Notice that Δ is an entire function of λ and $w = u, u_t$ [76], then we know that $\frac{\partial \lambda_j^{(c)}}{\partial w}$ is bounded, and

$$\frac{\partial F_j}{\partial w} = \left.\frac{\partial \Delta}{\partial w}\right|_{\lambda=\lambda_j^{(c)}}.$$ □

Theorem 4.6. *As a function of three variables,* $\Delta = \Delta(\lambda; u; u_t)$ *has the partial derivatives given by Bloch functions* $\psi^{\pm}$ *(i.e.,* $\psi^{\pm}(x) = e^{\pm \Lambda x}\tilde{\psi}^{\pm}(x)$, *where* $\tilde{\psi}^{\pm}$ *are periodic in* x *of period* 2π, *and* Λ *is a complex constant):*

$$\frac{\partial \Delta}{\partial u} = \frac{-i}{16\lambda c} \frac{\sqrt{\Delta^2 - 4}}{W(\psi^+, \psi^-)} (4\lambda c \partial_x(\psi_1^+\psi_2^- + \psi_2^+\psi_1^-) + e^{-iu}\psi_1^+\psi_1^- - e^{iu}\psi_2^+\psi_2^-),$$

$$\frac{\partial \Delta}{\partial u_t} = \frac{i}{4c} \frac{\sqrt{\Delta^2 - 4}}{W(\psi^+, \psi^-)} (\psi_1^+\psi_2^- + \psi_2^+\psi_1^-),$$

$$\frac{\partial \Delta}{\partial \lambda} = \frac{-1}{c} \frac{\sqrt{\Delta^2 - 4}}{W(\psi^+, \psi^-)} \int_0^{2\pi} \left[\left(\frac{1}{16\lambda^2}e^{iu} - 1\right)\psi_2^+\psi_2^- + \left(\frac{1}{16\lambda^2}e^{-iu} - 1\right)\psi_1^+\psi_1^-\right] dx,$$

where $W(\psi^+, \psi^-) = \psi_1^+\psi_2^- - \psi_2^+\psi_1^-$ *is the Wronskian.*

Proof: Recall that M is the fundamental matrix solution of (4.2), we have the equation for the differential of M

$$\partial_x dM = BdM + dBM, \quad dM(0) = 0.$$

Using the method of variation of parameters, we let

$$dM = MQ, \quad Q(0) = 0.$$

Thus

$$Q(x) = \int_0^x M^{-1} dBM dx,$$

and

$$dM(x) = M(x) \int_0^x M^{-1} dBM dx.$$

Finally

$$\begin{aligned} d\Delta &= \text{trace } (dM(2\pi)) \\ &= \text{trace } \left(M(2\pi) \int_0^{2\pi} M^{-1} dBM dx \right). \end{aligned} \tag{4.4}$$

Let

$$N = (\psi^+ \quad \psi^-),$$

where $\psi^\pm$ are two linearly independent Bloch functions (For the case that there is only one linearly independent Bloch function, L'Hospital's rule has to be used, for details, see [76]), such that

$$\psi^\pm = e^{\pm \Lambda x} \tilde{\psi}^\pm,$$

where $\tilde{\psi}^\pm$ are periodic in x of period 2π and Λ is a complex constant (The existence of such functions is the result of the well known Floquet theorem). Then

$$N(x) = M(x)N(0), \quad M(x) = N(x)[N(0)]^{-1}.$$

Notice that

$$N(2\pi) = N(0)E, \quad \text{where } E = \begin{pmatrix} e^{\Lambda 2\pi} & 0 \\ 0 & e^{-\Lambda 2\pi} \end{pmatrix}.$$

Then

$$M(2\pi) = N(0)E[N(0)]^{-1}.$$

Thus

$$\Delta = \text{ trace } M(2\pi) = \text{ trace } E = e^{\Lambda 2\pi} + e^{-\Lambda 2\pi},$$

and

$$e^{\pm \Lambda 2\pi} = \frac{1}{2}(\Delta \pm \sqrt{\Delta^2 - 4}).$$

In terms of N, $d\Delta$ as given in (4.4) takes the form

$$d\Delta = \text{trace}\left\{N(0)E[N(0)]^{-1}\int_0^{2\pi} N(0)[N(x)]^{-1}dB(x)N(x)[N(0)]^{-1}dx\right\}$$
$$= \text{trace}\left\{E\int_0^{2\pi}[N(x)]^{-1}dB(x)N(x)dx\right\},$$

from which one obtains the partial derivatives of Δ as stated in the theorem. □

It turns out that the partial derivatives of F_j provide the perfect Melnikov vectors rather than those of the Hamiltonian or other invariants [76], in the sense that F_j is the invariant whose level sets are the separatrices.

4.5 Heteroclinic Cycle

In this section, we will focus upon an example starting from $u = 0$. Linearizing the sine-Gordon equation at $u = 0$ leads to

$$u_{tt} = c^2 u_{xx} + u.$$

Let $u = \sum_{k=0}^{\infty} u_k(t)\cos(kx)$, then

$$\frac{d^2}{dt^2}u_k = (1 - c^2k^2)u_k.$$

Let $u_k \sim e^{\Omega_k t}$, then

$$\Omega_k = \pm\sqrt{1 - c^2k^2}, \quad k = 0, 1, 2, \ldots.$$

Since $1/2 < c < 1$, there are only two unstable modes $k = 0$ and 1. The corresponding nonlinear unstable foliation can be represented through the Darboux transformations given in Theorems 4.2 and 4.3. When $u = 0$, the Bloch functions of the Lax pair (4.2) and (4.3) are

$$\psi^{\pm} = e^{\pm i(\kappa x + \omega t)}\begin{pmatrix} 1 \\ \pm i \end{pmatrix},$$

where $\kappa = \frac{1}{c}(\lambda + \frac{1}{16\lambda})$ and $\omega = \lambda - \frac{1}{16\lambda}$. Thus,

$$\lambda = \frac{1}{2}\left[\kappa c \pm \sqrt{(\kappa c)^2 - \frac{1}{4}}\right].$$

The Floquet discriminant is given by

$$\Delta = 2\cos(2\pi\kappa). \tag{4.5}$$

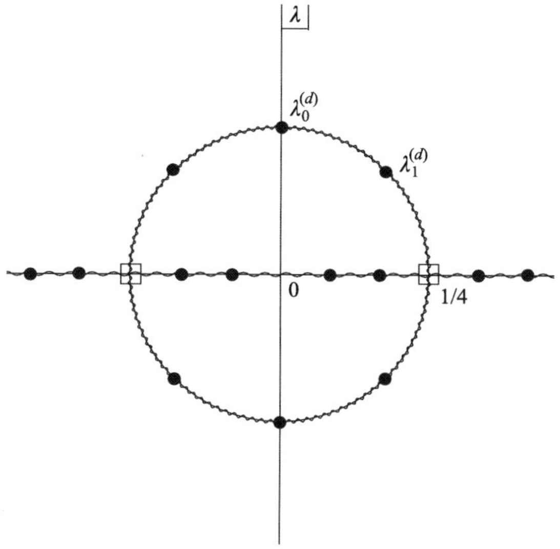

Figure 1 Floquet spectrum of the Lax pair at $u = 0$, • double point, □ critical point.

The spectral data are depicted in Figure 1. First we choose $\nu_0 = \lambda_0^{(d)} = \frac{i}{4}$ and use Theorem 4.2 to generate the foliation corresponding to $k = 0$. The corresponding Bloch functions are

$$\phi^\pm = \exp\left(\mp\frac{1}{2}t\right)\begin{pmatrix}1\\ \pm i\end{pmatrix}.$$

The wise choice for ϕ is

$$\phi = \sqrt{\frac{c^+}{c^-}}\phi^+ + \sqrt{\frac{c^-}{c^+}}\phi^-,$$

where $c^\pm$ are arbitrary complex constants. Let

$$\frac{c^+}{c^-} = e^{\rho + i\theta}, \quad \tau = t - \rho, \tag{4.6}$$

then

$$\phi = 2\begin{pmatrix}\cosh\frac{\tau}{2}\cos\frac{\theta}{2} - i\sinh\frac{\tau}{2}\sin\frac{\theta}{2}\\ -\cosh\frac{\tau}{2}\sin\frac{\theta}{2} - i\sinh\frac{\tau}{2}\cos\frac{\theta}{2}\end{pmatrix}. \tag{4.7}$$

By Theorem 4.2, we have

$$U = 2i\ln\left(i\frac{\phi_2}{\phi_1}\right),$$

where

$$i\frac{\phi_2}{\phi_1} = \frac{\tanh\tau - i\sin\theta\ \mathrm{sech}\ \tau}{1 + \cos\theta\ \mathrm{sech}\ \tau}.$$

For U to be real-valued, $\theta = \pm\pi/2$; and we have

$$U_{\pm} = \pm 2\vartheta, \tag{4.8}$$

where

$$\vartheta = \arccos \tanh \tau, \quad \vartheta \in [0, \pi].$$

In order to obtain the full foliation for the two unstable modes $k = 0$ and 1, one needs to apply the Darboux transformation II in Theorem 4.3 at $\nu_1 = \lambda_1^{(d)} = \frac{1}{4}[c + i\sqrt{1-c^2}]$ and $u = U_{\pm}$. At $(\lambda_1^{(d)}, u = 0)$, the Bloch functions are

$$\varphi^{\pm} = \exp\left(\pm i \frac{1}{2} x \mp \frac{\sigma}{2} t\right) \begin{pmatrix} 1 \\ \pm i \end{pmatrix}, \quad \sigma = \sqrt{1-c^2}, \tag{4.9}$$

Let

$$\varphi = \sqrt{\frac{c_+}{c_-}} \varphi^+ + \sqrt{\frac{c_-}{c_+}} \varphi^-,$$

where $c_{\pm}$ are arbitrary complex constants. Let

$$\frac{c_+}{c_-} = e^{\hat{\rho} + i\hat{\theta}}, \quad \hat{\tau} = \sigma t - \hat{\rho}, \quad \xi = x + \hat{\theta}; \tag{4.10}$$

then

$$\varphi = 2 \begin{pmatrix} \cosh \frac{\hat{\tau}}{2} \cos \frac{\xi}{2} - i \sinh \frac{\hat{\tau}}{2} \sin \frac{\xi}{2} \\ -\cosh \frac{\hat{\tau}}{2} \sin \frac{\xi}{2} - i \sinh \frac{\hat{\tau}}{2} \cos \frac{\xi}{2} \end{pmatrix}.$$

Applying the Darboux transformation in Theorem 4.2 to φ, one gets

$$\Phi = \begin{pmatrix} -\nu_0 \phi_2/\phi_1 & \nu_1 \\ -\nu_1 & \nu_0 \phi_1/\phi_2 \end{pmatrix} \varphi = \begin{pmatrix} i\nu_0 e^{\mp i\vartheta} \varphi_1 + \nu_1 \varphi_2 \\ -\nu_1 \varphi_1 + i\nu_0 e^{\pm i\vartheta} \varphi_2 \end{pmatrix}, \tag{4.11}$$

which solves the Lax pair at $(\nu_1, U_{\pm})$. Applying the Darboux transformation II in Theorem 4.3 at $(\nu_1, U_{\pm})$ with Φ, we have

$$U = \pm 2\vartheta - 4\hat{\vartheta},$$

where ϑ is given in (4.8), and $\hat{\vartheta} \in (-\pi/2, \pi/2)$,

$$\hat{\vartheta} = \arctan \frac{\sigma(\tanh \tau \ \mathrm{sech}\ \hat{\tau} \sin \xi \mp \ \mathrm{sech}\ \tau \tanh \hat{\tau})}{1 - \sigma(\tanh \tau \tanh \hat{\tau} \pm \ \mathrm{sech}\ \tau \ \mathrm{sech}\ \hat{\tau} \sin \xi)}.$$

For $\hat{\vartheta}$ to be even in x, we need to choose $\hat{\theta} = \pi/2$ or $-\pi/2$, and obtain

$$U_{\pm,\delta} = \pm 2\vartheta - 4\hat{\vartheta}, \quad \delta = + \text{ or } -;$$

where

$$\hat{\vartheta} = \arctan \frac{\sigma(\delta \tanh \tau \ \mathrm{sech}\ \hat{\tau} \cos x \mp \ \mathrm{sech}\ \tau \tanh \hat{\tau})}{1 - \sigma(\tanh \tau \tanh \hat{\tau} \pm \delta \ \mathrm{sech}\ \tau \ \mathrm{sech}\ \hat{\tau} \cos x)}.$$

The heteroclinic cycle is given by

$$\begin{aligned} U_1 &= U_{+,\delta} \\ &= 2\arccos(\tanh\tau) - 4\arctan\frac{\sigma(\pm\tanh\tau\ \text{sech}\ \hat{\tau}\cos x - \ \text{sech}\ \tau\tanh\hat{\tau})}{1-\sigma(\tanh\tau\tanh\hat{\tau}\pm\ \text{sech}\ \tau\ \text{sech}\ \hat{\tau}\cos x)}, \end{aligned} \tag{4.12}$$

$$\begin{aligned} U_2 &= 2\pi + U_{-,\delta} \\ &= 2\pi - 2\arccos(\tanh\tau) - 4\arctan\frac{\sigma(\pm\tanh\tau\ \text{sech}\ \hat{\tau}\cos x + \ \text{sech}\ \tau\tanh\hat{\tau})}{1-\sigma(\tanh\tau\tanh\hat{\tau}\mp\ \text{sech}\ \tau\ \text{sech}\ \hat{\tau}\cos x)}, \end{aligned} \tag{4.13}$$

where the ranges of arccos and arctan are $[0,\pi]$ and $(-\pi/2,\pi/2)$, and $(\tau,\sigma,\hat{\tau})$ are defined in (4.6), (4.9), and (4.10).

4.6 Melnikov Vectors Along the Heteroclinic Cycle

Since the sine-Gordon equation is invariant under the transformations $u \to u+2\pi$ and $u \to -u$, we have the following symmetries of the Lax pair (4.2) and (4.3).

Lemma 4.7. *If $\psi = (\psi_1,\psi_2)^T$ solves the Lax pair (4.2) and (4.3) at (λ,u), then it also solves the Lax pair (4.2) and (4.3) at $(\lambda,u+2\pi)$, and $(\psi_2,-\psi_1)^T$ solves the Lax pair (4.2) and (4.3) at $(\lambda,-u)$.*

Next we calculate the Melnikov vectors which will be given by $\frac{\partial F_j}{\partial u_t}$ $(j=0,1)$ where $\lambda_j^{(c)} = \nu_j$. By Lemma 4.7, the Melnikov vectors along U_1 and U_2 are the same as along $U_{+,\delta}$ and $U_{-,\delta}$. Expressions of the Melnikov vectors are given in Theorem 4.6. According to Theorem 4.2, let

$$\begin{aligned} \tilde{\phi}^{\pm} &= \begin{pmatrix} -\nu_0\phi_2/\phi_1 & \nu_0 \\ -\nu_0 & \nu_0\phi_1/\phi_2 \end{pmatrix}\phi^{\pm} \\ &= \mp\nu_0\exp\left(\mp\frac{\rho}{2}\mp i\frac{\theta}{2}\right)W(\phi^+,\phi^-)\begin{pmatrix} 1/\phi_1 \\ 1/\phi_2 \end{pmatrix}, \end{aligned}$$

where $\theta = \pi/2$ or $-\pi/2$. $\tilde{\phi}^{\pm}$ solves the Lax pair at (ν_0,U_+) or (ν_0,U_-) depending upon the choice of θ. According to Theorem 4.3, let

$$\tilde{\Phi}^{\pm} = \begin{pmatrix} \tilde{G}_1 & \tilde{G}_2 \\ \tilde{G}_3 & \tilde{G}_4 \end{pmatrix}\tilde{\phi}^{\pm}, \tag{4.14}$$

where

$$\tilde{G}_1 = |\nu_1|^2 \frac{\nu_1|\Phi_1|^2 + \overline{\nu_1}|\Phi_2|^2}{\overline{\nu_1}|\Phi_1|^2 + \nu_1|\Phi_2|^2} - \nu_0^2,$$
$$\tilde{G}_2 = \frac{\nu_0(\nu_1^2 - \overline{\nu_1}^2)\Phi_1\overline{\Phi_2}}{\overline{\nu_1}|\Phi_1|^2 + \nu_1|\Phi_2|^2},$$
$$\tilde{G}_3 = \frac{\nu_0(\nu_1^2 - \overline{\nu_1}^2)\overline{\Phi_1}\Phi_2}{\nu_1|\Phi_1|^2 + \overline{\nu_1}|\Phi_2|^2},$$
$$\tilde{G}_4 = |\nu_1|^2 \frac{\overline{\nu_1}|\Phi_1|^2 + \nu_1|\Phi_2|^2}{\nu_1|\Phi_1|^2 + \overline{\nu_1}|\Phi_2|^2} - \nu_0^2.$$

$\tilde{\Phi}^{\pm}$ solves the Lax pair at $(\nu_0, U_{\pm,\delta})$ where $\delta = +$ or $-$ depending on the choice of $\hat{\theta} = \pi/2$ or $-\pi/2$.

Similarly, according to Theorem 4.2, let

$$\Phi^{\pm} = \begin{pmatrix} -\nu_0\phi_2/\phi_1 & \nu_1 \\ -\nu_1 & \nu_0\phi_1/\phi_2 \end{pmatrix} \varphi^{\pm},$$

then

$$\Phi = \sqrt{\frac{c_+}{c_-}}\Phi^+ + \sqrt{\frac{c_-}{c_+}}\Phi^-.$$

$\Phi^{\pm}$ solves the Lax pair at (ν_1, U_+) or (ν_1, U_-) depending on the choice of $\theta = \pi/2$ or $-\pi/2$. According to Theorem 4.3, let

$$\hat{\Phi}^{\pm} = \begin{pmatrix} \hat{G}_1 & \hat{G}_2 \\ \hat{G}_3 & \hat{G}_4 \end{pmatrix} \Phi^{\pm},$$

where

$$\hat{G}_1 = \frac{\nu_1(\overline{\nu_1}^2 - \nu_1^2)|\Phi_2|^2}{\overline{\nu_1}|\Phi_1|^2 + \nu_1|\Phi_2|^2},$$
$$\hat{G}_2 = \frac{\nu_1(\nu_1^2 - \overline{\nu_1}^2)\Phi_1\overline{\Phi_2}}{\overline{\nu_1}|\Phi_1|^2 + \nu_1|\Phi_2|^2},$$
$$\hat{G}_3 = \frac{\nu_1(\nu_1^2 - \overline{\nu_1}^2)\overline{\Phi_1}\Phi_2}{\nu_1|\Phi_1|^2 + \overline{\nu_1}|\Phi_2|^2},$$
$$\hat{G}_4 = \frac{\nu_1(\overline{\nu_1}^2 - \nu_1^2)|\Phi_1|^2}{\nu_1|\Phi_1|^2 + \overline{\nu_1}|\Phi_2|^2}.$$

Then

$$\hat{\Phi}^{\pm} = \pm\nu_1(\nu_1^2 - \overline{\nu_1}^2)\exp\left(\mp\frac{\hat{\rho}}{2} \mp i\frac{\hat{\theta}}{2}\right) W(\Phi^+, \Phi^-) \begin{pmatrix} \dfrac{-\overline{\Phi_2}}{\overline{\nu_1}|\Phi_1|^2 + \nu_1|\Phi_2|^2} \\ \dfrac{\overline{\Phi_1}}{\nu_1|\Phi_1|^2 + \overline{\nu_1}|\Phi_2|^2} \end{pmatrix}. \quad (4.15)$$

$\hat{\Phi}^{\pm}$ solves the Lax pair at $(\nu_1, U_{\pm,\delta})$ where $\delta = +$ or $-$ depending on the choice of $\hat{\theta} = \pi/2$ or $-\pi/2$. Recall the process of obtaining (4.14) and (4.15), let $\psi^{\pm}$ be a pair of Bloch functions solving the Lax pair at $(\lambda, u = 0)$, then by Theorems 4.2 and 4.3,

$$\psi^{\pm} \xrightarrow{\nu_0} \tilde{\psi}^{\pm} \xrightarrow{\nu_1} \hat{\psi}^{\pm}.$$

Notice that in Theorem 4.3,

$$\det G = (\lambda^2 - \nu^2)(\lambda^2 - \bar{\nu}^2).$$

Thus

$$\begin{aligned} W(\hat{\psi}^+, \hat{\psi}^-) &= (\lambda^2 - \nu_1^2)(\lambda^2 - \overline{\nu_1}^2) W(\tilde{\psi}^+, \tilde{\psi}^-) \\ &= (\lambda^2 - \nu_1^2)(\lambda^2 - \overline{\nu_1}^2)(\lambda^2 - \nu_0^2) W(\psi^+, \psi^-). \end{aligned}$$

By the isospectral property, $U_{\pm,\delta}$, $U_{\pm}$ and $u = 0$ have the same Floquet discriminant $\Delta = \Delta(\lambda)$ as given in (4.5), $\Delta(\nu_0) = 2$, $\Delta(\nu_1) = -2$. In the neighborhood of ν_0 or ν_1, to the leading order,

$$\Delta^2 - 4 = (\Delta + 2)(\Delta - 2) = \Delta(\nu)\Delta''(\nu)(\lambda - \nu)^2, \quad \nu = \nu_0, \nu_1;$$

where

$$\Delta''(\nu) = -\frac{4\pi^2}{c^2\nu^2}(\nu - \bar{\nu})^2 \Delta(\nu).$$

Thus

$$\sqrt{\Delta^2 - 4} = i\frac{2\pi}{c\nu}(\nu - \bar{\nu})\Delta(\nu)(\lambda - \nu).$$

By L'Hospital's rule, as $\lambda \to \nu_0$

$$\frac{i}{4c}\frac{\sqrt{\Delta^2 - 4}}{W(\hat{\psi}^+, \hat{\psi}^-)} \to \frac{\pi(\overline{\nu_0} - \nu_0)}{2c^2\nu_0^2(\nu_0^2 - \nu_1^2)(\nu_0^2 - \overline{\nu_1}^2) W(\phi^+, \phi^-)}, \tag{4.16}$$

and as $\lambda \to \nu_1$

$$\frac{i}{4c}\frac{\sqrt{\Delta^2 - 4}}{W(\hat{\psi}^+, \hat{\psi}^-)} \to \frac{\pi}{2c^2\nu_1^2(\nu_1 + \overline{\nu_1})(\nu_1^2 - \nu_0^2) W(\varphi^+, \varphi^-)}. \tag{4.17}$$

Finally, from Theorem 4.6 and relations (4.14)–(4.17), we have

$$\begin{aligned} \frac{\partial F_0}{\partial u_t} &= \frac{\pi(\nu_0 - \overline{\nu_0}) W(\phi^+, \phi^-)}{c^2(\nu_0^2 - \nu_1^2)(\nu_0^2 - \overline{\nu_1}^2)} (\tilde{G}_1/\phi_1 + \tilde{G}_2/\phi_2)(\tilde{G}_3/\phi_1 + \tilde{G}_4/\phi_2) \\ &= \frac{64\pi}{c^4}(\tilde{G}_1/\phi_1 + \tilde{G}_2/\phi_2)(\tilde{G}_3/\phi_1 + \tilde{G}_4/\phi_2), \\ \frac{\partial F_1}{\partial u_t} &= \frac{\pi}{c^2}(\nu_1 - \overline{\nu_1})(\nu_1^2 - \overline{\nu_1}^2)(\nu_1^2 - \nu_0^2) W(\varphi^+, \varphi^-) \\ &\quad \cdot \frac{\overline{\Phi_1 \Phi_2}}{(\nu_1|\Phi_1|^2 + \overline{\nu_1}|\Phi_2|^2)(\overline{\nu_1}|\Phi_1|^2 + \nu_1|\Phi_2|^2)} \end{aligned} \tag{4.18}$$

$$= \frac{\pi(1-c^2)}{4c} i(\nu_1^2 - \nu_0^2) \frac{\overline{\Phi_1 \Phi_2}}{(\nu_1|\Phi_1|^2 + \overline{\nu_1}|\Phi_2|^2)(\overline{\nu_1}|\Phi_1|^2 + \nu_1|\Phi_2|^2)}, \quad (4.19)$$

where ϕ is given in (4.7) with $\theta = \pm\pi/2$, $\tilde{G}$ is given in (4.14), and Φ is given in (4.11).

Chapter 5

Heisenberg Ferromagnet Equation

5.1 Background

The Heisenberg model is a model in statistical physics to model ferromagnetism. The continuum limit of the Heisenberg model leads to the Heisenberg ferromagnet equation. It extends to the more realistic ferromagnet equations like the Landau-Lifshitz equation and Landau-Lifshitz-Gilbert equation when other effects of the ferromagnet are taken into account.

The Heisenberg ferromagnet equation is another important integrable system [34]. Although there exists a correspondence between the Heisenberg ferromagnet equation and the integrable nonlinear Schrödinger equation, this correspondence does not render the two problems equivalent. We start with the Heisenberg ferromagnet equation in the form

$$\partial_t m = -m \times m_{xx}, \tag{5.1}$$

subject to the periodic boundary condition

$$m(t, x + 2\pi) = m(t, x), \tag{5.2}$$

where $m = (m_1, M_2, m_3)^T$ and $|m|(t, x) = 1$. Introduce the matrix

$$\Gamma = m_j \sigma_j = \begin{pmatrix} m_3 & m_- \\ m_+ & -m_3 \end{pmatrix}, \tag{5.3}$$

where σ_j are the Pauli matrices

$$\sigma_1 = \begin{pmatrix} 0 & 1 \\ 1 & 0 \end{pmatrix}, \quad \sigma_2 = \begin{pmatrix} 0 & -i \\ i & 0 \end{pmatrix}, \quad \sigma_3 = \begin{pmatrix} 1 & 0 \\ 0 & -1 \end{pmatrix}, \tag{5.4}$$

and

$$m_+ = m_1 + im_2, \quad m_- = m_1 - im_2, \tag{5.5}$$

i.e., $\overline{m_+} = m_-$. Using the matrix Γ introduced in (5.3), the Heisenberg equation (5.1) has the form

$$i\partial_t \Gamma = -\frac{1}{2}[\Gamma, \Gamma_{xx}], \tag{5.6}$$

where $[\Gamma, \Gamma_{xx}] = \Gamma\Gamma_{xx} - \Gamma_{xx}\Gamma$.

5.2 Lax Pair

The Heisenberg equation (5.1) is an integrable system with the following Lax pair,

$$\partial_x \psi = i\lambda\Gamma\psi, \tag{5.7}$$

$$\partial_t \psi = -\frac{\lambda}{2}\left(4i\lambda\Gamma + [\Gamma, \Gamma_x]\right)\psi, \tag{5.8}$$

where $\psi = (\psi_1, \psi_2)^T$ is complex-valued, λ is a complex parameter, Γ is the matrix defined in (5.3), and $[\Gamma, \Gamma_x] = \Gamma\Gamma_x - \Gamma_x\Gamma$. For more details on this chapter, see [56].

5.3 Darboux Transformations

A Darboux transformation for (5.7) and (5.8) can be obtained.

Theorem 5.1. *Let $\phi = (\phi_1, \phi_2)^T$ be a solution to the Lax pair (5.7) and (5.8) at (Γ, ν). Define the matrix*

$$G = N \begin{pmatrix} (\nu - \lambda)/\nu & 0 \\ 0 & (\bar{\nu} - \lambda)/\bar{\nu} \end{pmatrix} N^{-1},$$

where

$$N = \begin{pmatrix} \phi_1 & -\overline{\phi_2} \\ \phi_2 & \overline{\phi_1} \end{pmatrix}.$$

Then if ψ solves the Lax pair (5.7) and (5.8) at (Γ, λ),

$$\hat{\psi} = G\psi \tag{5.9}$$

solves the Lax pair (5.7) and (5.8) at $(\hat{\Gamma}, \lambda)$, where $\hat{\Gamma}$ is given by

$$\hat{\Gamma} = N \begin{pmatrix} e^{-i\theta} & 0 \\ 0 & e^{i\theta} \end{pmatrix} N^{-1} \Gamma N \begin{pmatrix} e^{i\theta} & 0 \\ 0 & e^{-i\theta} \end{pmatrix} N^{-1}, \tag{5.10}$$

where $e^{i\theta} = \nu/|\nu|$.

The transformations (5.9) and (5.10) are the Darboux transformation. This theorem can be proved either through the connection between the Heisenberg equation and the NLS equation (with a well-known Darboux transformation) [23], or through a direct calculation.

Notice also that $\hat{\Gamma}^2 = I$. Let

$$\begin{pmatrix} \Phi_1 & -\overline{\Phi_2} \\ \Phi_2 & \overline{\Phi_1} \end{pmatrix} = N \begin{pmatrix} e^{-i\theta} & 0 \\ 0 & e^{i\theta} \end{pmatrix} N^{-1}$$

$$= \frac{1}{|\phi_1|^2 + |\phi_2|^2} \begin{pmatrix} e^{-i\theta}|\phi_1|^2 + e^{i\theta}|\phi_2|^2 & -2i\sin\theta\ \phi_1\overline{\phi_2} \\ -2i\sin\theta\ \overline{\phi_1}\phi_2 & e^{i\theta}|\phi_1|^2 + e^{-i\theta}|\phi_2|^2 \end{pmatrix}. \quad (5.11)$$

Then

$$\hat{\Gamma} = \begin{pmatrix} \hat{m}_3 & \hat{m}_1 - i\hat{m}_2 \\ \hat{m}_1 + i\hat{m}_2 & -\hat{m}_3 \end{pmatrix}, \quad (5.12)$$

where

$$\hat{m}_+ = \hat{m}_1 + i\hat{m}_2 = \overline{\Phi_1}^2(m_1 + im_2) - \Phi_2^2(m_1 - im_2) + 2\overline{\Phi_1}\Phi_2 m_3,$$

$$\hat{m}_3 = \left(|\Phi_1|^2 - |\Phi_2|^2\right) m_3 - \overline{\Phi_1\Phi_2}(m_1 + im_2) - \Phi_1\Phi_2(m_1 - im_2).$$

One can generate the figure eight connecting to the domain wall, as the nonlinear foliation of the modulational instability, via the above Darboux transformation.

5.4 Figure Eight Structures Connecting to the Domain Wall

Let Γ be the domain wall

$$\Gamma = \begin{pmatrix} 0 & e^{-i2x} \\ e^{i2x} & 0 \end{pmatrix},$$

i.e., $m_1 = \cos 2x$, $m_2 = \sin 2x$, and $m_3 = 0$. Solving the Lax pair (5.7) and (5.8), one gets two Bloch eigenfunctions

$$\psi = e^{\Omega t} \begin{pmatrix} 2\lambda \exp\left(\frac{i}{2}(k-2)x\right) \\ (k-2)\exp\left(\frac{i}{2}(k+2)x\right) \end{pmatrix}, \quad \Omega = -i\lambda k, \quad k = \pm 2\sqrt{1+\lambda^2}. \quad (5.13)$$

To apply the Darboux transformation (5.10), we start with the two Bloch functions with $k = \pm 1$,

$$\phi^+ = \begin{pmatrix} \sqrt{3}e^{-ix} \\ ie^{ix} \end{pmatrix} \exp\left(\frac{\sqrt{3}}{2}t + i\frac{1}{2}x\right),$$

$$\phi^- = \begin{pmatrix} -ie^{-ix} \\ \sqrt{3}e^{ix} \end{pmatrix} \exp\left(-\frac{\sqrt{3}}{2}t - i\frac{1}{2}x\right). \quad (5.14)$$

The wise choice for ϕ used in (5.10) is:

$$\phi = \sqrt{\frac{c^+}{c^-}}\phi^+ + \sqrt{\frac{c^-}{c^+}}\phi^- = \begin{pmatrix} \left(\sqrt{3}e^{\tau+i\chi} - ie^{-\tau-i\chi}\right)e^{-ix} \\ \left(ie^{\tau+i\chi} + \sqrt{3}e^{-\tau-i\chi}\right)e^{ix} \end{pmatrix}, \tag{5.15}$$

where $c^+/c^- = \exp(\sigma + i\gamma)$, $\tau = \frac{1}{2}(\sqrt{3}t + \sigma)$, and $\chi = \frac{1}{2}(x + \gamma)$. Then from the Darboux transformation (5.10), one gets

$$\hat{m}_1 + i\hat{m}_2 = -e^{i2x}\left\{1 - \frac{2 \text{ sech } 2\tau \cos 2\chi}{(2 - \sqrt{3} \text{ sech } 2\tau \sin 2\chi)^2}[\text{sech } 2\tau \cos 2\chi \right. \tag{5.16}$$

$$\left. + i(\sqrt{3} - 2 \text{ sech } 2\tau \sin 2\chi)]\right\},$$

$$\hat{m}_3 = \frac{2 \text{ sech } 2\tau \tanh 2\tau \cos 2\chi}{(2 - \sqrt{3} \text{ sech } 2\tau \sin 2\chi)^2}. \tag{5.17}$$

As $t \to \pm\infty$,

$$\hat{m}_1 \to -\cos 2x, \quad \hat{m}_2 \to -\sin 2x, \quad \hat{m}_3 \to 0.$$

The expressions (5.16) and (5.17) represent the two dimensional figure eight separatrix connecting to the domain wall ($m_+ = -e^{i2x}$, $m_3 = 0$), parametrized by σ and γ. See Figure 1 for an illustration. Choosing $\gamma = 0, \pi$, one gets the figure eight curve section of Figure 1. The spatial-temporal profiles corresponding to the two lobes of the figure eight curve are shown in Figure 2. In fact, the two profiles corresponding the two lobes are spatial translates of each other by π. Inside one of the lobe, the spatial-temporal profile is shown in Figure 3(a). Outside the figure eight curve, the spatial-temporal profile is shown in Figure 3(b). Here the inside and outside spatial-temporal profiles are calculated by using the integrable finite difference discretization [50] of the Heisenberg equation (5.1),

$$\frac{d}{dt}m(j) = -\frac{2}{h^2}m(j) \times \left[\frac{m(j+1)}{1 + m(j)\cdot m(j+1)} + \frac{m(j-1)}{1 + m(j-1)\cdot m(j)}\right], \tag{5.18}$$

where $m(j) = m(t, jh)$, $j = 1, \ldots, N$, $Nh = 2\pi$, and h is the spatial mesh size. For the computation of Figure 3, we choose $N = 128$.

By a translation $x \to x + \theta$, one can generate a circle of domain walls:

$$m_+ = -e^{i2(x+\theta)}, \quad m_3 = 0,$$

where θ is the phase parameter. The three dimensional figure eight separatrix connecting to the circle of domain walls, parametrized by σ, γ and θ, is illustrated in Figure 4.

In general, the unimodal equilibrium manifold can be sought as follows: Let

$$m_j = c_j \cos 2x + s_j \sin 2x, \quad j = 1, 2, 3,$$

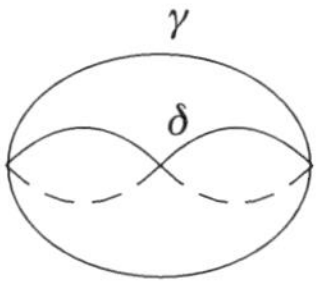

Figure 1 The separatrix connecting to the domain wall $m_+ = -e^{i2x}, m_3 = 0$.

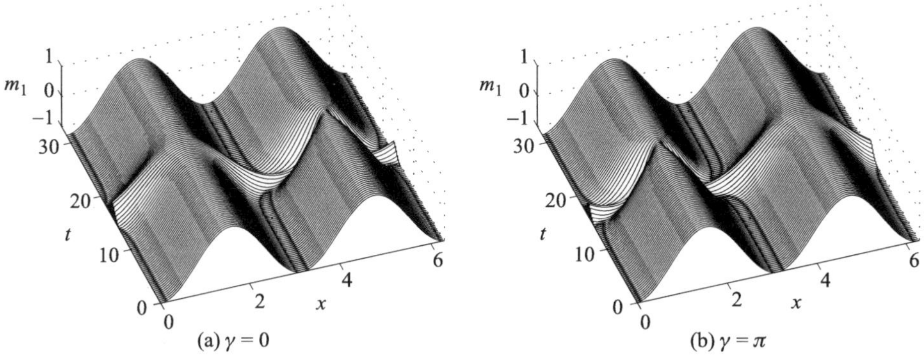

Figure 2 The spatial-temporal profiles corresponding to the two lobes of the figure eight curve.

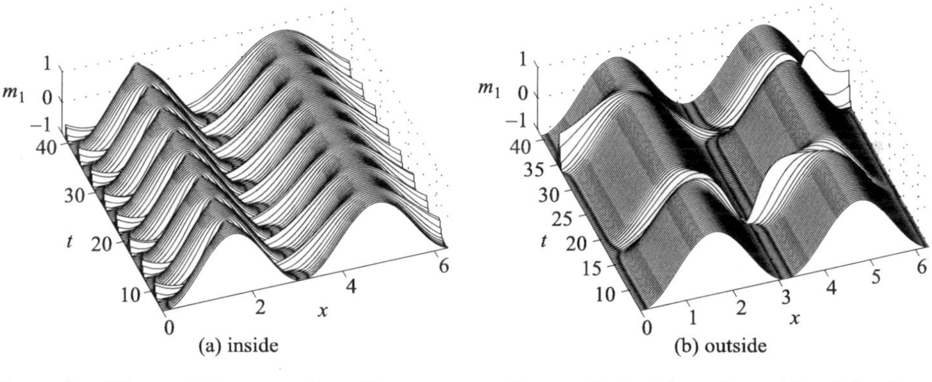

Figure 3 The spatial-temporal profiles corresponding to the inside and outside of the figure eight curve.

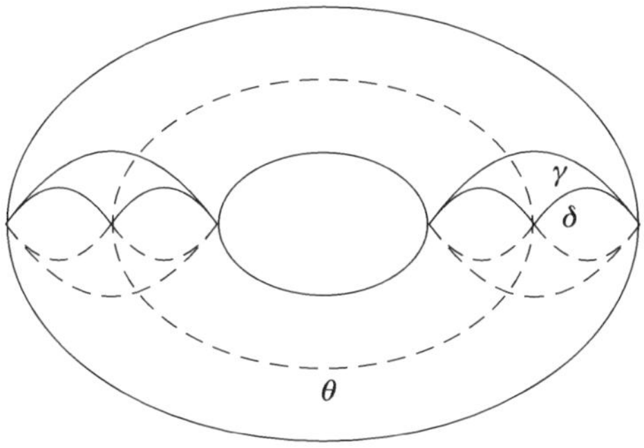

Figure 4 The separatrix connecting to the circle of domain walls $m_+ = e^{i2(x+\theta)}, m_3 = 0$.

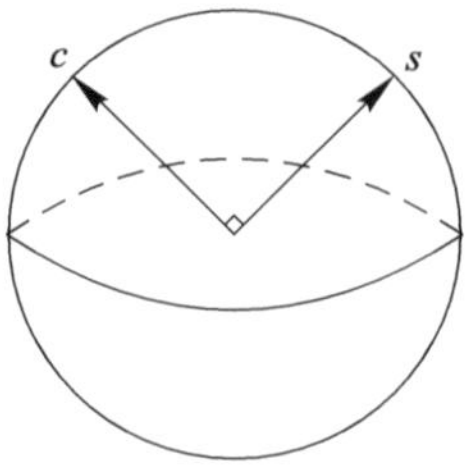

Figure 5 A representation of the 3 dimensional unimodal equilibrium manifold.

then the uni-length condition $|m|(x) = 1$ leads to

$$|c| = 1, \quad |s| = 1, \quad c \cdot s = 0,$$

where c and s are the two vectors with components c_j and s_j. Thus the unimodal equilibrium manifold is three dimensional and can be represented as in Figure 5.

5.5 Floquet Theory

Focusing on the spatial part (5.7) of the Lax pair (5.7) and (5.8), let $Y(x)$ be the fundamental matrix solution of (5.7), $Y(0) = I$ (2×2 identity matrix), then the Floquet discriminant is defined by

$$\Delta = \text{ trace } Y(2\pi).$$

The Floquet spectrum is given by

$$\sigma = \{\lambda \in \mathbb{C} | -2 \leq \Delta(\lambda) \leq 2\}.$$

Periodic and anti-periodic points $\lambda^{\pm}$ (which correspond to periodic and anti-periodic eigenfunctions respectively) are defined by

$$\Delta(\lambda^{\pm}) = \pm 2.$$

A critical point $\lambda^{(c)}$ is defined by

$$\frac{d\Delta}{d\lambda}(\lambda^{(c)}) = 0.$$

A multiple point $\lambda^{(n)}$ is a periodic or anti-periodic point which is also a critical point. The algebraic multiplicity of $\lambda^{(n)}$ is defined as the order of the zero of $\Delta(\lambda) \pm 2$ at $\lambda^{(n)}$. When the order is 2, we call the multiple point a double point, and denote it by $\lambda^{(d)}$. The order can exceed 2. The geometric multiplicity of $\lambda^{(n)}$ is defined as the dimension of the periodic or anti-periodic eigenspace at $\lambda^{(n)}$, and is either 1 or 2.

Counting lemmas for $\lambda^{\pm}$ and $\lambda^{(c)}$ can be established as in [76, 90], which lead to the existence of the sequences $\{\lambda_j^{\pm}\}$ and $\{\lambda_j^{(c)}\}$ and their approximate locations. For any $\lambda \in \mathbb{C}$, $\Delta(\lambda)$ is a constant of motion of the Heisenberg equation (5.1).

Example 5.2. For the domain wall $m_1 = \cos 2x$, $m_2 = \sin 2x$, and $m_3 = 0$; the two Bloch eigenfunctions are given in (5.13). The Floquet discriminant is given by

$$\Delta = 2\cos(2\pi\sqrt{1+\lambda^2}).$$

The periodic points are given by

$$\lambda = \pm\sqrt{\frac{n^2}{4} - 1}, \quad n \in \mathbb{Z}, \quad n \text{ is even}.$$

The anti-periodic points are given by

$$\lambda = \pm\sqrt{\frac{n^2}{4} - 1}, \quad n \in \mathbb{Z}, \quad n \text{ is odd}.$$

The choice of ϕ^+ and ϕ^- correspond to $n = \pm 1$ and $\lambda = \nu = i\sqrt{3}/2$ with $k = \pm 1$.

$$\Delta' = -4\pi \frac{\lambda}{\sqrt{1+\lambda^2}} \sin(2\pi\sqrt{1+\lambda^2}).$$

$$\Delta'' = -4\pi(1+\lambda^2)^{-3/2}\sin(2\pi\sqrt{1+\lambda^2}) - 8\pi^2\frac{\lambda^2}{1+\lambda^2}\cos(2\pi\sqrt{1+\lambda^2}).$$

When $n = 0$, i.e., $\sqrt{1+\lambda^2} = 0$, by L'Hospital's rule

$$\Delta' \to -8\pi^2\lambda, \quad \lambda = \pm i.$$

That is, $\lambda = \pm i$ are periodic points, not critical points. When $n = \pm 1$, we have two imaginary double points

$$\lambda = \pm i\sqrt{3}/2.$$

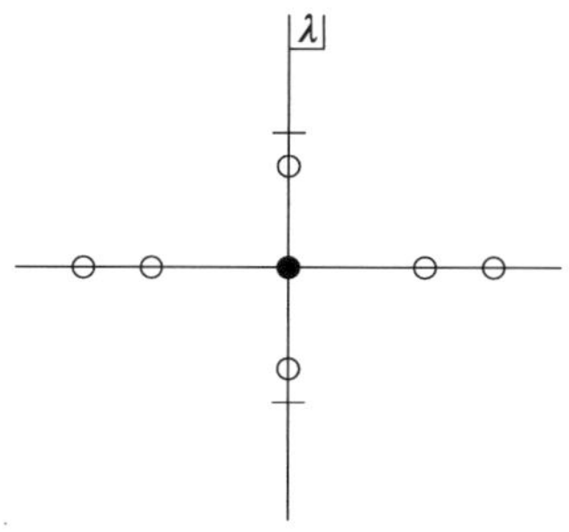

Figure 6 The periodic and anti-periodic points corresponding to the potential of domain wall $m_+ = e^{i2x}$, $m_3 = 0$. The open circles are double points, the solid circle at the origin is a multiple point of order 4, and the two bars intersect the imaginary axis at two periodic points which are not critical points.

When $n = \pm 2$, $\lambda = 0$ is a multiple point of order 4. The rest periodic and anti-periodic points are all real double points. Figure 6 is an illustration of these spectral points.

5.6 Melnikov Vectors

Starting from the Floquet theory, one can build Melnikov vectors.

Definition 5.3. An important sequence of invariants F_j of the Heisenberg equation is defined by

$$F_j(m) = \Delta(\lambda_j^{(c)}(m), m).$$

Lemma 5.4. *If* $\{\lambda_j^{(c)}\}$ *is a simple critical point of* Δ, *then*

$$\frac{\partial F_j}{\partial m} = \left.\frac{\partial \Delta}{\partial m}\right|_{\lambda=\lambda_j^{(c)}}.$$

Proof: We know that

$$\frac{\partial F_j}{\partial m} = \left.\frac{\partial \Delta}{\partial m}\right|_{\lambda=\lambda_j^{(c)}} + \left.\frac{\partial \Delta}{\partial \lambda}\right|_{\lambda=\lambda_j^{(c)}} \frac{\partial \lambda_j^{(c)}}{\partial m}.$$

Since

$$\left.\frac{\partial \Delta}{\partial \lambda}\right|_{\lambda=\lambda_j^{(c)}} = 0,$$

we have

$$\left.\frac{\partial^2 \Delta}{\partial \lambda^2}\right|_{\lambda=\lambda_j^{(c)}} \frac{\partial \lambda_j^{(c)}}{\partial m} + \left.\frac{\partial^2 \Delta}{\partial \lambda \partial m}\right|_{\lambda=\lambda_j^{(c)}} = 0.$$

Since $\lambda_j^{(c)}$ is a simple critical point of Δ,

$$\left.\frac{\partial^2 \Delta}{\partial \lambda^2}\right|_{\lambda=\lambda_j^{(c)}} \neq 0.$$

Thus

$$\frac{\partial \lambda_j^{(c)}}{\partial m} = -\left(\left.\frac{\partial^2 \Delta}{\partial \lambda^2}\right|_{\lambda=\lambda_j^{(c)}}\right)^{-1} \left.\frac{\partial^2 \Delta}{\partial \lambda \partial m}\right|_{\lambda=\lambda_j^{(c)}}.$$

Notice that Δ is an entire function of λ and m [76], then we know that $\frac{\partial \lambda_j^{(c)}}{\partial m}$ is bounded, and

$$\frac{\partial F_j}{\partial m} = \left.\frac{\partial \Delta}{\partial m}\right|_{\lambda=\lambda_j^{(c)}}.$$

□

Theorem 5.5. *As a function of two variables, $\Delta = \Delta(\lambda, m)$ has the partial derivatives given by Bloch functions $\psi^\pm$ (i.e., $\psi^\pm(x) = e^{\pm\Lambda x}\tilde{\psi}^\pm(x)$, where $\tilde{\psi}^\pm$ are periodic in x of period 2π, and Λ is a complex constant):*

$$\frac{\partial\Delta}{\partial m_+} = -i\lambda\frac{\sqrt{\Delta^2-4}}{W(\psi^+,\psi^-)}\psi_1^+\psi_1^-,$$

$$\frac{\partial\Delta}{\partial m_-} = i\lambda\frac{\sqrt{\Delta^2-4}}{W(\psi^+,\psi^-)}\psi_2^+\psi_2^-,$$

$$\frac{\partial\Delta}{\partial m_3} = i\lambda\frac{\sqrt{\Delta^2-4}}{W(\psi^+,\psi^-)}\left(\psi_1^+\psi_2^- + \psi_2^+\psi_1^-\right),$$

$$\frac{\partial\Delta}{\partial \lambda} = i\frac{\sqrt{\Delta^2-4}}{W(\psi^+,\psi^-)}\int_0^{2\pi}\left[m_3\left(\psi_1^+\psi_2^- + \psi_2^+\psi_1^-\right) - m_+\psi_1^+\psi_1^- + m_-\psi_2^+\psi_2^-\right]dx,$$

where $W(\psi^+,\psi^-) = \psi_1^+\psi_2^- - \psi_2^+\psi_1^-$ is the Wronskian.

Proof: Recall that Y is the fundamental matrix solution of (5.7), we have the equation for the differential of Y

$$\partial_x dY = i\lambda\Gamma dY + i(d\lambda\Gamma + \lambda d\Gamma)Y, \quad dY(0) = 0.$$

Using the method of variation of parameters, we let

$$dY = YQ, \quad Q(0) = 0.$$

Thus

$$Q(x) = i\int_0^x Y^{-1}(d\lambda\Gamma + \lambda d\Gamma)Y\,dx,$$

and

$$dY(x) = iY\int_0^x Y^{-1}(d\lambda\Gamma + \lambda d\Gamma)Y\,dx.$$

Finally

$$\begin{aligned} d\Delta =& \text{ trace } dY(2\pi) \\ =& \; i \text{ trace } \left[Y(2\pi)\int_0^{2\pi} Y^{-1}(d\lambda\Gamma + \lambda d\Gamma)Y\,dx\right]. \end{aligned} \tag{5.19}$$

Let

$$Z = (\psi^+\psi^-)$$

where $\psi^\pm$ are two linearly independent Bloch functions (For the case that there is only one linearly independent Bloch function, L'Hospital's rule has to be used, for details, see [76]), such that

$$\psi^\pm = e^{\pm\Lambda x}\tilde{\psi}^\pm,$$

where $\tilde{\psi}^{\pm}$ are periodic in x of period 2π and Λ is a complex constant (The existence of such functions is the result of the well known Floquet theorem). Then

$$Z(x) = Y(x)Z(0), \quad Y(x) = Z(x)[Z(0)]^{-1}.$$

Notice that

$$Z(2\pi) = Z(0)E, \quad \text{where } E = \begin{pmatrix} e^{\Lambda 2\pi} & 0 \\ 0 & e^{-\Lambda 2\pi} \end{pmatrix}.$$

Then

$$Y(2\pi) = Z(0)E[Z(0)]^{-1}.$$

Thus

$$\Delta = \text{ trace } Y(2\pi) = \text{ trace } E = e^{\Lambda 2\pi} + e^{-\Lambda 2\pi},$$

and

$$e^{\pm \Lambda 2\pi} = \frac{1}{2}[\Delta \pm \sqrt{\Delta^2 - 4}].$$

In terms of Z, $d\Delta$ as given in (5.19) takes the form

$$\begin{aligned} d\Delta &= i \text{ trace } \left\{ Z(0)E[Z(0)]^{-1} \int_0^{2\pi} Z(0)[Z(x)]^{-1}(d\lambda\Gamma + \lambda d\Gamma)Z(x)[Z(0)]^{-1}dx \right\} \\ &= i \text{ trace } \left\{ E \int_0^{2\pi} [Z(x)]^{-1}(d\lambda\Gamma + \lambda d\Gamma)Z(x)dx \right\}, \end{aligned}$$

from which one obtains the partial derivatives of Δ as stated in the theorem. □

It turns out that the partial derivatives of F_j provide the perfect Melnikov vectors rather than those of the Hamiltonian or other invariants [76], in the sense that F_j is the invariant whose level sets are the separatrices.

5.7 Melnikov Vectors Along the Figure Eight Structures

We continue the calculation in Subsection 5.4 to obtain an explicit expression of the Melnikov vector along the figure eight connecting to the domain wall. Apply the Darboux transformation (5.9) to $\phi^{\pm}$ (5.14) at $\lambda = \nu$, we obtain

$$\hat{\phi}^{\pm} = \pm \frac{\bar{\nu} - \nu}{\bar{\nu}} \frac{\exp(\mp\frac{1}{2}\sigma \mp i\frac{1}{2}\gamma)W(\phi^+, \phi^-)}{|\phi_1|^2 + |\phi_2|^2} \begin{pmatrix} \overline{\phi_2} \\ -\overline{\phi_1} \end{pmatrix}.$$

In the formula (5.9), for general λ,

$$\det G = \frac{(\nu - \lambda)(\bar{\nu} - \lambda)}{|\nu|^2}, \quad W(\hat{\psi}^+, \hat{\psi}^-) = \det G \, W(\psi^+, \psi^-).$$

In a neighborhood of $\lambda = \nu$,

$$\Delta^2 - 4 = \Delta(\nu)\Delta''(\nu)(\lambda - \nu)^2 + \text{ higher order terms in } (\lambda - \nu).$$

As $\lambda \to \nu$, by L'Hospital's rule

$$\frac{\sqrt{\Delta^2 - 4}}{W(\hat{\psi}^+, \hat{\psi}^-)} \to \frac{\sqrt{\Delta(\nu)\Delta''(\nu)}}{\frac{\nu - \bar{\nu}}{|\nu|^2} W(\phi^+, \phi^-)}.$$

Notice, by the calculation in Example 5.2, that

$$\nu = i\frac{\sqrt{3}}{2}, \quad \Delta(\nu) = -2, \quad \Delta''(\nu) = -24\pi^2,$$

then by Theorem 5.5,

$$\begin{aligned}
\left.\frac{\partial \Delta}{\partial m_+}\right|_{m=\hat{m}} &= 12\sqrt{3}\pi \frac{i}{(|\phi_1|^2 + |\phi_2|^2)^2} \overline{\phi_2}^{\,2}, \\
\left.\frac{\partial \Delta}{\partial m_-}\right|_{m=\hat{m}} &= 12\sqrt{3}\pi \frac{-i}{(|\phi_1|^2 + |\phi_2|^2)^2} \overline{\phi_1}^{\,2}, \\
\left.\frac{\partial \Delta}{\partial m_3}\right|_{m=\hat{m}} &= 12\sqrt{3}\pi \frac{2i}{(|\phi_1|^2 + |\phi_2|^2)^2} \overline{\phi_1 \phi_2},
\end{aligned}$$

where $\hat{m}$ is given in (5.16) and (5.17). With the explicit expression (5.15) of ϕ, we obtain the explicit expressions of the Melnikov vector,

$$\begin{aligned}
\left.\frac{\partial \Delta}{\partial m_+}\right|_{m=\hat{m}} &= \frac{3\sqrt{3}\pi}{2} \frac{i \text{ sech } 2\tau}{(2 - \sqrt{3} \text{ sech } 2\tau \sin 2\chi)^2} [(1 - 2\tanh\ 2\tau)\cos 2\chi \qquad (5.20) \\
&\quad + i(2 - \tanh 2\tau)\sin 2\chi - i\sqrt{3} \text{ sech } 2\tau] e^{-i2x},
\end{aligned}$$

$$\begin{aligned}
\left.\frac{\partial \Delta}{\partial m_-}\right|_{m=\hat{m}} &= \frac{3\sqrt{3}\pi}{2} \frac{-i \text{ sech } 2\tau}{(2 - \sqrt{3} \text{ sech } 2\tau \sin 2\chi)^2} [(1 + 2\tanh 2\tau)\cos 2\chi \qquad (5.21) \\
&\quad - i(2 + \tanh 2\tau)\sin 2\chi + i\sqrt{3} \text{ sech } 2\tau] e^{i2x},
\end{aligned}$$

$$\begin{aligned}
\left.\frac{\partial \Delta}{\partial m_3}\right|_{m=\hat{m}} &= \frac{3\sqrt{3}\pi}{2} \frac{2i \text{ sech } 2\tau}{(2 - \sqrt{3} \text{ sech } 2\tau \sin 2\chi)^2} [2 \text{ sech } 2\tau - \sqrt{3}\sin 2\chi \qquad (5.22) \\
&\quad - i\sqrt{3}\tanh 2\tau \cos 2\chi],
\end{aligned}$$

where again

$$m_\pm = m_1 \pm i m_2, \quad \tau = \frac{\sqrt{3}}{2}t + \frac{\sigma}{2}, \quad \chi = \frac{1}{2}(x + \gamma),$$

and σ and γ are two real parameters.

5.8 A Melnikov Function for Landau-Lifshitz-Gilbert Equation

Discovered by Slonczewski [99] and Berger [17], electrical current can apply a large torque to a ferromagnet. If electrical current can be used to achieve magnetization reversal, such a nanotechnique will dramatically increase the magnetic memory capacity and speed. This subject has just recently emerged as one of the most important subjects in science and technology [51, 106]. Taking into account of the torque induced by electrical current, the governing equation is given by the Landau-Lifshitz-Gilbert (LLG) equation which can be written in the dimensionless form as

$$\partial_t m = -m \times H - \alpha m \times (m \times H) + \beta m \times (m \times e_x), \tag{5.23}$$

subject to the periodic boundary condition (5.2), where $|m|(t,x) = 1$, the effective magnetic field H has several terms

$$\begin{aligned} H &= H_{\text{exch}} + H_{\text{ext}} + H_{\text{dem}} + H_{\text{ani}} \\ &= \partial_x^2 m + a e_x - b m_3 e_z + c m_1 e_x, \end{aligned} \tag{5.24}$$

where $H_{\text{exch}} = \partial_x^2 m$ is the exchange field, $H_{\text{ext}} = a e_x$ is the external field, $H_{\text{dem}} = -b m_3 e_z$ is the demagnetization field, and $H_{\text{ani}} = c m_1 e_x$ is the anisotropy field. For the materials of the experimental interest, the dimensionless parameters are in the ranges

$$\begin{aligned} &a \in [-0.05, 0.05], \quad b \in [0, 0.005], \quad c \in [0.05, 0.1], \\ &\alpha = [0.008, 0.02], \quad \beta \in [-0.08, 0.08]. \end{aligned} \tag{5.25}$$

The Landau-Lifshitz-Gilbert (LLG) equation (5.23) can be rewritten in the form,

$$\partial_t m = -m \times m_{xx} + f \tag{5.26}$$

where f is the perturbation

$$\begin{aligned} f = &-a m \times e_x + b m_3 (m \times e_z) - c m_1 (m \times e_x) \\ &-\alpha m \times (m \times H) + \beta m \times (m \times e_x), \end{aligned}$$

and

$$H = m_{xx} + a e_x - b m_3 e_z + c m_1 e_x.$$

The Melnikov function for the LLG (5.23) is given as

$$M = \int_{-\infty}^{\infty} \int_0^{2\pi} \left[\frac{\partial \Delta}{\partial m_+} (f_1 + i f_2) + \frac{\partial \Delta}{\partial m_-} (f_1 - i f_2) + \frac{\partial \Delta}{\partial m_3} f_3 \right] \Bigg|_{m = \hat{m}} dx dt,$$

where $\hat{m}$ is given in (5.16) and (5.17), and $\frac{\partial \Delta}{\partial w}$ $(w = m_+, m_-, m_3)$ are given in (5.20) and (5.22). The Melnikov function depends on several external and internal parameters $M = M(a, b, c, \alpha, \beta, \gamma)$ where γ is an internal parameter.

Chapter 6

Vector Nonlinear Schrödinger Equations

6.1 Physical Background

In birefringent optical fibers, optical signals have two polarization directions. The propagation of two optical pulse envelopes is described by the vector nonlinear Schrödinger equation [86, 87]

$$ip_t + p_{xx} + \frac{1}{2}(|p|^2 + \kappa|q|^2)p = 0,$$
$$iq_t + q_{xx} + \frac{1}{2}(\kappa|p|^2 + |q|^2)q = 0,$$

where p and q are complex-valued functions of (t, x), $\kappa \in [0, 1]$. When $\kappa = 0, 1$; it reduces to an integrable system. Otherwise, it is non-integrable. Optical fibers now have important industrial applications in fiber communication systems (see [42] for an introductory reference), and all-optical switching devices [48]. This chapter will focus on the $\kappa = 1$ case, for more details, see [64]. In general, the following n-vector nonlinear Schrödinger equation is integrable [2]

$$i\vec{q}_t + \vec{q}_{xx} + |\vec{q}|^2\vec{q} = 0,$$

where $\vec{q} = (q_1, \ldots, q_n)$.

In the parametric wave context [89], Faraday surface waves in a liquid container are generated by oscillatory vertical shaking of the container. The envelope equation of the weakly nonlinear Faraday waves is the vector Ginzburg-Landau equation which is closely connected to the vector nonlinear Schrödinger equation.

6.2 Lax Pair

Consider the integrable vector nonlinear Schrödinger equations (VNLS),

$$ip_t + p_{xx} + \frac{1}{2}[(|p|^2 + |q|^2) - \omega^2]p = 0, \quad (6.1)$$

$$iq_t + q_{xx} + \frac{1}{2}[(|p|^2 + |q|^2) - \omega^2]q = 0, \tag{6.2}$$

where $\omega \in (1, 2)$. Periodic boundary condition is imposed

$$p(t, x + 2\pi) = p(t, x), \quad q(t, x + 2\pi) = q(t, x).$$

The Lax pair is given as [83],

$$\psi_x = U\psi, \quad \psi_t = V\psi, \tag{6.3}$$

where

$$U = \lambda A_0 + A_1, \quad V = \lambda^2 A_0 + \lambda A_1 + A_2,$$

$$A_0 = \begin{pmatrix} -\frac{2}{3}i & 0 & 0 \\ 0 & \frac{1}{3}i & 0 \\ 0 & 0 & \frac{1}{3}i \end{pmatrix}, \quad A_1 = \begin{pmatrix} 0 & \frac{1}{2}p & \frac{1}{2}q \\ -\frac{1}{2}\bar{p} & 0 & 0 \\ -\frac{1}{2}\bar{q} & 0 & 0 \end{pmatrix},$$

$$A_2 = -\frac{i}{4}\begin{pmatrix} -(|p|^2 + |q|^2) + 4\omega^2 & -2p_x & -2q_x \\ -2\bar{p}_x & |p|^2 + 2\omega^2 & q\bar{p} \\ -2\bar{q}_x & \bar{q}p & |q|^2 + 2\omega^2 \end{pmatrix}.$$

Notice that here the spatial part of the Lax pair is a 3×3 problem for which the Floquet theory is more complicated than the 2×2 problems before. In particular,it is difficult to build Melnikov vectors via Floquet theory in this case.

6.3 Linearized Equations

We start with the solution,

$$p = ae^{i\delta_1}, \quad q = be^{i\delta_2}, \tag{6.4}$$

where

$$\delta_1 = \frac{1}{2}(a^2 + b^2 - \omega^2)t + \gamma_1, \quad \delta_2 = \frac{1}{2}(a^2 + b^2 - \omega^2)t + \gamma_2.$$

Linearizing (6.1) and (6.2) at this solution, by setting

$$p = e^{i\delta_1}[a + \tilde{p}], \quad q = e^{i\delta_2}[b + \tilde{q}],$$

one gets

$$i\tilde{p}_t + \tilde{p}_{xx} + \frac{1}{2}a[a(\tilde{p} + \bar{\tilde{p}}) + b(\tilde{q} + \bar{\tilde{q}})] = 0,$$

$$i\tilde{q}_t + \tilde{q}_{xx} + \frac{1}{2}b[a(\tilde{p} + \bar{\tilde{p}}) + b(\tilde{q} + \bar{\tilde{q}})] = 0.$$

Setting

$$\tilde{p} = f_+ e^{ikx+\Omega t} + \overline{f_-} e^{-ikx+\bar{\Omega} t}, \quad \tilde{q} = g_+ e^{ikx+\Omega t} + \overline{g_-} e^{-ikx+\bar{\Omega} t}, \quad (k \in Z)$$

one gets the dispersion relations

$$\Omega_{1,2} = \pm k\sqrt{a^2 + b^2 - k^2}, \quad \Omega_{3,4} = \pm i k^2. \tag{6.5}$$

Thus, if we choose $1 < \sqrt{a^2 + b^2} < 2$, then there is only one unstable mode $\cos x$. This is the reason that we restricts ω as in (6.1) and (6.2). When $k = 1$,

$$\Omega_{1,2} = \pm\sigma, \quad \sigma = \sqrt{a^2 + b^2 - 1}. \tag{6.6}$$

The corresponding eigenfunctions are

$$\tilde{p} = a e^{\pm i\varphi} e^{\pm\sigma t} \cos x, \quad \tilde{q} = b e^{\pm i\varphi} e^{\pm\sigma t} \cos x, \quad \varphi = \arctan \sigma.$$

6.4 Homoclinic Orbits and Figure Eight Structures

The Bäcklund-Darboux transformation given in [35, 107] will be utilized to construct homoclinic orbits asymptotic to the periodic orbits (6.4) up to phase shifts.

Theorem 6.1 ([35, 107]). *Let p and q be a solution of the vector nonlinear Schrödinger equations* (6.1) *and* (6.2), *and let $\psi = (\psi_1, \psi_2, \psi_3)^T$ be an eigenfunction of the Lax pair* (6.3) *at (λ, p, q). If one defines*

$$\hat{p} = p + 2i(\bar{\lambda} - \lambda)\frac{\psi_1\overline{\psi_2}}{|\psi_1|^2 + |\psi_2|^2 + |\psi_3|^2}, \tag{6.7}$$

$$\hat{q} = q + 2i(\bar{\lambda} - \lambda)\frac{\psi_1\overline{\psi_3}}{|\psi_1|^2 + |\psi_2|^2 + |\psi_3|^2}, \tag{6.8}$$

then $\hat{p}$ and $\hat{q}$ also solve the vector nonlinear Schrödinger equations (6.1) *and* (6.2).

We solve the Lax pair (6.3) at (λ, p, q) with p and q given by the periodic orbits (6.4), we get

$$\psi^\pm = \begin{pmatrix} \psi_1^\pm \\ \psi_2^\pm \\ \psi_3^\pm \end{pmatrix} = \begin{pmatrix} (i\lambda \mp i)e^{i(\delta_1+\delta_2)} \\ a e^{i\delta_2} \\ b e^{i\delta_1} \end{pmatrix} e^{\pm\sigma_1 t + i\sigma_2 t} e^{\pm i\kappa^\pm x}, \tag{6.9}$$

where

$$\kappa^\pm = -\frac{1}{6}\lambda \pm \frac{1}{2}\sqrt{a^2 + b^2 + \lambda^2}, \quad \sqrt{a^2 + b^2 + \lambda^2} = 1, \tag{6.10}$$

$$\sigma_1 = \frac{i\lambda}{2}, \quad \sigma_2 = -\frac{1}{12}[7(a^2 + b^2) + 2]. \tag{6.11}$$

Remark 6.2. From (6.11), we see that in order to have temporal growth, the imaginary part of λ can not be zero. Also in order to have a nontrivial Bäcklund-Darboux transformation (Theorem 6.1), the imaginary part of λ can not be zero. From (6.10), if $a^2+b^2 > 1$, then λ is purely imaginary. If we also require $a^2+b^2 < 4$, i.e., in the one unstable mode regime (6.5), then $\sqrt{a^2+b^2+\lambda^2} \geq 2$ will lead to real λ and no temporal growth.

Remark 6.3. Notice that the imaginary part of $\kappa^\pm$ is not zero, thus $\psi^\pm$ are not periodic or antiperiodic functions in x. In fact, they grow or decay in x. This fact shows that complex double points are not necessary in constructing homoclinic orbits through Bäcklund-Darboux transformations (see also [62]).

Let

$$\psi = c^+\psi^+ + c^-\psi^-, \tag{6.12}$$

where c^+ and c^- are two arbitrary complex constants. We introduce the notations

$$c^+/c^- = e^{\rho+i\vartheta}, \quad \lambda = -i\sigma, \quad \sigma = \sqrt{a^2+b^2-1}, \tag{6.13}$$

$$\sigma + i = \sqrt{a^2+b^2}e^{i(\frac{\pi}{2}-\vartheta_0)}, \quad \vartheta_0 = \arctan\sigma, \quad \sigma_1 = \sigma/2, \tag{6.14}$$

$$\tau = \frac{\sigma}{2}t + \frac{\rho}{2}, \quad y = \frac{1}{2}x + \frac{\vartheta}{2}, \tag{6.15}$$

where ρ and ϑ are called the Bäcklund parameters. If one chooses $\lambda = i\sigma$, one will get the same result. We can rewrite ψ in the following form,

$$\begin{aligned}\psi_1 &= \left[\cosh\tau\cos\left(y-\frac{\pi}{2}+\vartheta_0\right) + i\sinh\tau\sin\left(y-\frac{\pi}{2}+\vartheta_0\right)\right]\\ &\quad \cdot 2\sqrt{c^+c^-}\sqrt{a^2+b^2}e^{i(\delta_1+\delta_2)}e^{i\sigma_2 t}e^{-\frac{1}{6}\sigma x},\\ \psi_2 &= (\cosh\tau\cos y + i\sinh\tau\sin y)2\sqrt{c^+c^-}ae^{i\delta_2}e^{i\sigma_2 t}e^{-\frac{1}{6}\sigma x},\\ \psi_3 &= (\cosh\tau\cos y + i\sinh\tau\sin y)2\sqrt{c^+c^-}be^{i\delta_1}e^{i\sigma_2 t}e^{-\frac{1}{6}\sigma x},\end{aligned}$$

where the factor $2\sqrt{c^+c^-}e^{i\sigma_2 t}e^{-\frac{1}{6}\sigma x}$ will be canceled away. By (6.7) and (6.8), one gets

$$\hat{p} = ae^{i\delta_1}h, \quad \hat{q} = be^{i\delta_2}h, \tag{6.16}$$

where

$$\begin{aligned}h &= \left(\cos 2\vartheta_0 + i\sin 2\vartheta_0\tanh 2\tau - \sin\vartheta_0\ \mathrm{sech}\ 2\tau\cos\left(2y+\vartheta_0-\frac{\pi}{2}\right)\right)\\ &\quad \cdot\left(1+\sin\vartheta_0\ \mathrm{sech}\ 2\tau\cos\left(2y+\vartheta_0-\frac{\pi}{2}\right)\right)^{-1},\end{aligned} \tag{6.17}$$

where δ_1 and δ_2 are given in (6.4), other notations are given in (6.13)—(6.15). For the moment, the even-in-x condition has not been enforced. For fixed a and b, the phases γ_1 and γ_2 in (6.4), and the Bäcklund parameters ρ and ϑ parametrize a four dimensional submanifold with a figure eight structure. If $a = b$ and $\gamma_1 = \gamma_2$, then the homoclinic orbit reduces to that for the scalar nonlinear Schrödinger equation [70].

As $t \to \pm\infty$,

$$h \to e^{\pm i2\vartheta_0}. \tag{6.18}$$

The even-in-x condition can be achieved by restricting the Bäcklund parameter ϑ to special values. That is, if one requires that

$$\vartheta + \vartheta_0 - \frac{\pi}{2} = 0,\ \pi, \tag{6.19}$$

then h is even in x,

$$\begin{aligned} h = &(\cos 2\vartheta_0 + i \sin 2\vartheta_0 \tanh 2\tau \mp \sin\vartheta_0 \text{ sech } 2\tau \cos x) \\ &\cdot(1 \pm \sin\vartheta_0 \text{ sech } 2\tau \cos x)^{-1}, \end{aligned} \tag{6.20}$$

where the upper sign corresponds to $\vartheta + \vartheta_0 - \frac{\pi}{2} = 0$. For fixed a and b, the phases γ_1 and γ_2 in (6.4), and the Bäcklund parameter ρ parametrize a three dimensional submanifold with a figure eight structure. If one further chooses $\gamma_1 = \gamma_2$ in (6.4), then (6.4) represents a periodic orbit. In such case, the phase $\gamma_1 = \gamma_2$ and the Bäcklund parameter ρ parametrize a two dimensional submanifold with a figure eight structure.

6.5 A Melnikov Vector

The Hamiltonian for the integrable vector nonlinear Schrödinger equations (6.1) and (6.2) is given as,

$$H = \int_0^{2\pi} \left\{ |p_x|^2 + |q_x|^2 - \frac{1}{4}[(|p|^2 + |q|^2)^2 - 2\omega^2(|p|^2 + |q|^2)] \right\} dx. \tag{6.21}$$

In fact, (6.1) and (6.2) can be rewritten in the Hamiltonian form,

$$ip_t = \frac{\delta H}{\delta \bar{p}}, \quad i\bar{p}_t = -\frac{\delta H}{\delta p}, \quad iq_t = \frac{\delta H}{\delta \bar{q}}, \quad i\bar{q}_t = -\frac{\delta H}{\delta q}.$$

The two L^2 norms

$$E_1 = \int_0^{2\pi} |p|^2 dx, \quad E_2 = \int_0^{2\pi} |q|^2 dx, \tag{6.22}$$

are also invariant functionals. We can build the Melnikov vector by the gradient of the invariant functional G obtained through a combination out of H, E_1, and E_2,

$$\begin{aligned} G &= H + \frac{1}{2}\left[\frac{1}{4\pi}(E_1 + E_2) - \omega^2\right](E_1 + E_2) \\ &= \int_0^{2\pi} \left[|p_x|^2 + |q_x|^2 - \frac{1}{4}(|p|^2 + |q|^2)^2\right] dx + \frac{1}{8\pi}(E_1 + E_2)^2. \end{aligned} \tag{6.23}$$

As mentioned before, because of the fact that the spatial part of the Lax pair (6.3) is a 3×3 system, it is difficult to build the Melnikov vector out of the Floquet theory. The above Melnikov vector offers an alternative.

Chapter 7

Derivative Nonlinear Schrödinger Equations

7.1 Physical Background

In plasma physics, the propagation of Alfvén waves is described by the derivative nonlinear Schrödinger equation. It also describes the propagation of ultra-short optical pulses in optical fibers, the propagation of electromagnet waves in ferromagnetism, and many other physical phenomena [100]. Several forms of the derivative nonlinear Schrödinger equations are integrable, e.g., [33]

$$iq_t = q_{xx} + i\alpha(|q|^2 q)_x + 2|q|^2 q, \quad \alpha \text{ is a real parameter,}$$

and [74, 100]

$$iq_t = q_{xx} - i(|q|^2 q)_x.$$

In this chapter, we shall focus on the second one. For more details on this chapter, see [74].

7.2 Lax Pair

Here we study the derivative nonlinear Schrödinger equation (DNLS) in the form,

$$iq_t = q_{xx} - i\left(|q|^2 q\right)_x, \tag{7.1}$$

where q is a complex-valued function of two variables t and x. Periodic boundary condition is imposed:

$$q(t, x + 2\pi) = q(t, x).$$

DNLS is an integrable system, and its Lax pair is given as

$$\partial_x \psi = U\psi, \tag{7.2}$$

$$\partial_t \psi = V\psi, \tag{7.3}$$

where

$$U = \begin{pmatrix} -i\lambda^2 & \lambda q \\ \lambda \bar{q} & i\lambda^2 \end{pmatrix},$$

$$V = \begin{pmatrix} i2\lambda^4 + i\lambda^2|q|^2 & -2\lambda^3 q - \lambda(iq_x + |q|^2 q) \\ -2\lambda^3\bar{q} + \lambda(i\bar{q}_x - |q|^2\bar{q}) & -i2\lambda^4 - i\lambda^2|q|^2 \end{pmatrix}.$$

The Lax pair possesses some symmetries as stated in the following lemma.

Lemma 7.1. *If $\psi = (\psi_1, \psi_2)^T$ solves the Lax pair (7.2) and (7.3) at (λ, q), then $(\overline{\psi_2}, \overline{\psi_1})^T$ solves the Lax pair at $(\bar{\lambda}, q)$ and $(\psi_1, -\psi_2)^T$ solves the Lax pair at $(-\lambda, q)$.*

7.3 Darboux Transformations

There are several forms of the Darboux transformation for the DNLS. Here we adopt the form obtained in [100].

Theorem 7.2 (Darboux Transformation I). *Let q be a solution of the DNLS (7.1), let $\phi = (\phi_1, \phi_2)^T$ be a solution to the Lax pair (7.2) and (7.3) at $\lambda = \nu$ for some ν. Define the matrix*

$$G = \begin{pmatrix} \lambda\phi_2/\phi_1 & -\nu \\ -\nu & \lambda\phi_1/\phi_2 \end{pmatrix}.$$

Let

$$Q = \frac{\phi_2^2}{\phi_1^2} q - 2i\nu\frac{\phi_2}{\phi_1}, \qquad \Psi = G\psi;$$

then Q is also a solution of the DNLS, and Ψ solves the Lax pair at (λ, Q); provided that

$$\bar{Q} = \frac{\phi_1^2}{\phi_2^2}\bar{q} + 2i\nu\frac{\phi_1}{\phi_2}. \tag{7.4}$$

The Darboux transformation in this theorem does not guarantee reality condition (7.4). In some cases, for instance ν is real and $|\phi_1| = |\phi_2|$, reality condition (7.4) can be achieved. It is such a case that leads us to the strange tori to be discussed later. When ν is complex, an iteration of the Darboux transformation in Theorem 7.2 leads to a new form of the Darboux transformation which guarantees reality condition.

Theorem 7.3 (Darboux Transformation II). *Let q be a solution of the DNLS (7.1), let $\phi = (\phi_1, \phi_2)^T$ be a solution to the Lax pair (7.2) and (7.3) at $\lambda = \nu$ for some ν (ν complex). Define the matrix*

$$G = \begin{pmatrix} G_{11} & G_{12} \\ G_{21} & G_{22} \end{pmatrix},$$

where

$$G_{11} = \lambda^2 \frac{\bar{\nu}|\phi_1|^2 - \nu|\phi_2|^2}{\bar{\nu}|\phi_2|^2 - \nu|\phi_1|^2} + |\nu|^2,$$

$$G_{12} = \frac{\lambda(\nu^2 - \bar{\nu}^2)\phi_1\overline{\phi_2}}{\bar{\nu}|\phi_2|^2 - \nu|\phi_1|^2},$$

$$G_{21} = \frac{\lambda(\nu^2 - \bar{\nu}^2)\phi_2\overline{\phi_1}}{\bar{\nu}|\phi_1|^2 - \nu|\phi_2|^2},$$

$$G_{22} = \lambda^2 \frac{\bar{\nu}|\phi_2|^2 - \nu|\phi_1|^2}{\bar{\nu}|\phi_1|^2 - \nu|\phi_2|^2} + |\nu|^2.$$

Let

$$Q = \frac{\bar{\nu}|\phi_1|^2 - \nu|\phi_2|^2}{\bar{\nu}|\phi_2|^2 - \nu|\phi_1|^2} \frac{\bar{\nu}|\phi_1|^2 - \nu|\phi_2|^2}{\bar{\nu}|\phi_2|^2 - \nu|\phi_1|^2} q + \frac{2i(\nu^2 - \bar{\nu}^2)\phi_1\overline{\phi_2}}{\bar{\nu}|\phi_2|^2 - \nu|\phi_1|^2},$$

and

$$\Psi = G\psi.$$

Then Q is also a solution of the DNLS, and Ψ solves the Lax pair at (λ, Q).

Proof: Since ϕ solves the Lax pair at $\lambda = \nu$, by Lemma 7.1, $(\overline{\phi_2}, \overline{\phi_1})^T$ solves the Lax pair at $\lambda = \bar{\nu}$. Iterating the Darboux transformation in Theorem 7.2 at $\bar{\nu}$, we have

$$\Phi = \begin{pmatrix} \Phi_1 \\ \Phi_2 \end{pmatrix} = \begin{pmatrix} \bar{\nu}\phi_2/\phi_1 & -\nu \\ -\nu & \bar{\nu}\phi_1/\phi_2 \end{pmatrix} \begin{pmatrix} \overline{\phi_2} \\ \overline{\phi_1} \end{pmatrix}$$

$$= \begin{pmatrix} (\bar{\nu}|\phi_2|^2 - \nu|\phi_1|^2)/\phi_1 \\ (\bar{\nu}|\phi_1|^2 - \nu|\phi_2|^2)/\phi_2 \end{pmatrix},$$

$$\hat{G} = \begin{pmatrix} \lambda\Phi_2/\Phi_1 & -\bar{\nu} \\ -\bar{\nu} & \lambda\Phi_1/\Phi_2 \end{pmatrix}.$$

Then

$$\tilde{\Psi} = \hat{G}\Psi = \tilde{G}\psi,$$

where

$$\tilde{G} = \hat{G}G = \begin{pmatrix} \tilde{G}_{11} & \tilde{G}_{12} \\ \tilde{G}_{21} & \tilde{G}_{22} \end{pmatrix},$$

with $\tilde{G}_{mn}$ $(m, n = 1, 2)$ are given as in the Theorem 7.3 (without $\tilde{}$). Similarly

$$\tilde{Q} = \frac{\Phi_2^2}{\Phi_1^2} Q - 2i\bar{\nu}\frac{\Phi_2}{\Phi_1},$$

which leads to the expression given as in the Theorem 7.3 (without $\tilde{}$). The corresponding transform of (7.4) is

$$\frac{\Phi_1^2}{\Phi_2^2}\left(\frac{\phi_1^2}{\phi_2^2}\bar{q} + 2i\nu\frac{\phi_1}{\phi_2}\right) + 2i\bar{\nu}\frac{\Phi_1}{\Phi_2},$$

which can be verified directly to be just $\bar{\tilde{Q}}$. □

7.4 Floquet Theory

Focusing upon the spatial part (7.2) of the Lax pair, one can develop a complete Floquet theory. Let $M(x)$ be the fundamental matrix of (7.2), $M(0) = I$ (2×2 identity matrix), then the Floquet discriminant is given as

$$\Delta = \text{ trace } M(2\pi).$$

The Floquet spectrum is given by

$$\sigma = \{\lambda \in \mathbb{C} | -2 \leq \Delta(\lambda) \leq 2\}.$$

Periodic and anti-periodic points $\lambda^{\pm}$ (which correspond to periodic and anti-periodic eigenfunctions respectively) are defined by

$$\Delta(\lambda^{\pm}) = \pm 2.$$

A critical point $\lambda^{(c)}$ is defined by

$$\frac{d\Delta}{d\lambda}(\lambda^{(c)}) = 0.$$

A multiple point $\lambda^{(m)}$ is a periodic or anti-periodic point which is also a critical point. The algebraic multiplicity of $\lambda^{(m)}$ is defined as the order of the zero of $\Delta(\lambda) \pm 2$ at $\lambda^{(m)}$. When the order is 2, we call the multiple point a double point, and denote it by $\lambda^{(d)}$. The order can exceed 2. The geometric multiplicity of $\lambda^{(m)}$ is defined as the dimension of the periodic or anti-periodic eigenspace at $\lambda^{(m)}$, and is either 1 or 2.

Counting lemmas for $\lambda^{\pm}$ and $\lambda^{(c)}$ can be established as in [76, 90], which lead to the existence of the sequences $\{\lambda_j^{\pm}\}$ and $\{\lambda_j^{(c)}\}$ and their approximate locations. For any $\lambda \in \mathbb{C}$, $\Delta(\lambda)$ is invariant under the DNLS flow.

7.5 Strange Tori

For the monochromatic wave solution

$$q = ae^{i\vartheta}, \quad \vartheta = kx + \omega t + \theta, \quad \omega = k(k - a^2), \tag{7.5}$$

where $k \in \mathbb{Z}$, $a > 0$ and $\theta \in [0, 2\pi]$; the eigenfunctions of the Lax pair (7.2) and (7.3) can be calculated,

$$\psi^{\pm} = \begin{pmatrix} \lambda a e^{i\vartheta/2} \\ i\left(\lambda^2 + \frac{k}{2} + \xi_{\pm}\right) e^{-i\vartheta/2} \end{pmatrix} \exp(i\xi_{\pm}x + \Omega_{\pm}t), \tag{7.6}$$

where

$$\xi_\pm = \pm\sqrt{\left(\lambda^2 + \frac{k}{2}\right)^2 - \lambda^2 a^2}, \quad \Omega_\pm = i\xi_\pm(k - a^2 - 2\lambda^2).$$

The Floquet discriminant is given as

$$\Delta = 2\cos\left[2\pi\left(\xi_+ + \frac{k}{2}\right)\right] = 2\cos\left[2\pi\sqrt{\left(\lambda^2 + \frac{k}{2}\right)^2 - \lambda^2 a^2} + k\pi\right].$$

From now on, I choose $k = 2$ and $(a^2 - 2)^2 < 3$, i.e.,

$$\sqrt{2 - \sqrt{3}} < a < \sqrt{2 + \sqrt{3}},$$

so that there are only two linearly unstable modes at the monochromatic wave. In this case,

$$\Delta = 2\cos\left[2\pi\sqrt{(\lambda^2 + 1)^2 - \lambda^2 a^2}\right],$$

and

$$\Delta' = 4\pi\lambda\frac{2(\lambda^2 + 1) - a^2}{\sqrt{(\lambda^2 + 1)^2 - \lambda^2 a^2}}\sin\left[2\pi\sqrt{(\lambda^2 + 1)^2 - \lambda^2 a^2}\right].$$

Periodic and antiperiodic points are given by

$$\lambda^2 = \frac{1}{2}(a^2 - 2) \pm \frac{1}{2}\sqrt{(a^2 - 2)^2 + (n^2 - 4)} \tag{7.7}$$

which correspond to $\xi_+ = \frac{n}{2}$ $(n \geq 0, n \in \mathbb{Z})$. Except for $n = 0$, all the rest periodic or antiperiodic points are also double points. See Figure 1 for an illustration.

Let $\xi_+ = n/2$, $n \geq 2$, $n \in \mathbb{Z}$. Then pick $\lambda = \nu_n > 0$ where

$$\nu_n^2 = \frac{1}{2}(a^2 - 2) + \frac{1}{2}\sqrt{(a^2 - 2)^2 + (n^2 - 4)}.$$

Let

$$\begin{aligned} \phi &= c^+\phi^+ + c^-\phi^- \\ &= \begin{pmatrix} \nu_n a e^{i\vartheta/2}\left[c^+ e^{i\frac{n}{2}x + \tilde{\Omega}_+ t} + c^- e^{-i\frac{n}{2}x + \tilde{\Omega}_- t}\right] \\ ie^{-i\vartheta/2}\left[c^+\left(\nu_n^2 + 1 + \frac{n}{2}\right)e^{i\frac{n}{2}x + \tilde{\Omega}_+ t} + c^-\left(\nu_n^2 + 1 - \frac{n}{2}\right)e^{-i\frac{n}{2}x + \tilde{\Omega}_- t}\right] \end{pmatrix}, \end{aligned}$$

where $\tilde{\Omega}_\pm = \pm i\frac{n}{2}(2 - a^2 - 2\nu_n^2)$. To apply the Darboux transformation in Theorem 7.2, we need to have $|\phi_1| = |\phi_2|$. This leads to the condition

$$c^+/c^- = \sqrt{\frac{\nu_n^2 + 1 - \frac{n}{2}}{\nu_n^2 + 1 + \frac{n}{2}}}e^{i\gamma}$$

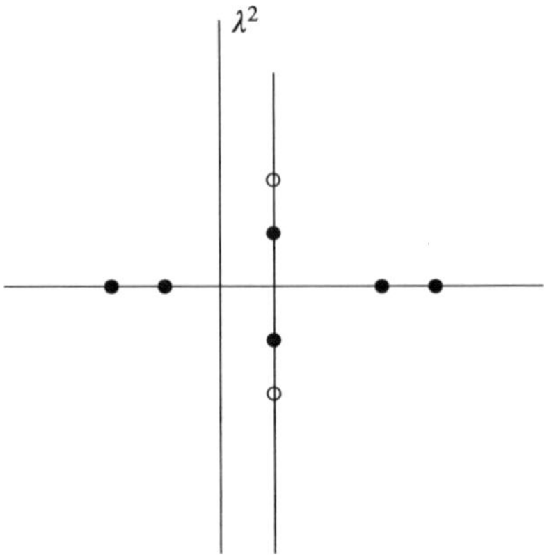

Figure 1 Periodic and antiperiodic spectra.

where γ is a phase parameter, $\gamma \in [0, 2\pi]$. In this case,

$$\phi_2/\phi_1 = ie^{-i\vartheta} \frac{\sqrt{\nu_n^2 + 1 + \frac{n}{2}} e^{inx + 2\tilde{\Omega}_+ t + i\gamma} + \sqrt{\nu_n^2 + 1 - \frac{n}{2}}}{\sqrt{\nu_n^2 + 1 - \frac{n}{2}} e^{inx + 2\tilde{\Omega}_+ t + i\gamma} + \sqrt{\nu_n^2 + 1 + \frac{n}{2}}}, \tag{7.8}$$

and

$$Q = \frac{\phi_2}{\phi_1} \left(\frac{\phi_2}{\phi_1} q - 2i\nu_n \right), \tag{7.9}$$

which in general represents a 2-torus $\mathbb{T}^2$. The two frequencies are

$$\omega = 2(2 - a^2), \quad \omega_n = -i2\tilde{\Omega}_+ = n \left[2(2 - a^2) - \sqrt{(a^2 - 2)^2 + (n^2 - 4)} \right].$$

The strangeness of the $\mathbb{T}^2$ comes from the fact that the Floquet discriminants for q and Q are identical

$$\Delta(\lambda, q) = \Delta(\lambda, Q). \tag{7.10}$$

That is, the $\mathbb{T}^2$ represented by Q and the circle represented by q lie on the same level set determined by $\Delta(\lambda)$ for all $\lambda \in \mathbb{C}$. The $\mathbb{T}^2$ and the circle are obviously disconnected pieces. For Q, the geometric multiplicity at $\lambda = \nu_n$ is 1; while it is 2 for q. The algebraic multiplicity at $\lambda = \nu_n$ for both Q and q is 2. Such a multiplicity feature is also true for a torus and its whisker. This was well understood before [76].

To see the relation (7.10), one notices that ϕ_2/ϕ_1 in (7.8) is periodic in x, thus the matrix G in Theorem 7.2 is periodic in x, and Q and q have the same Floquet discriminant.

Of course, Q can reduce to a periodic orbit or a fixed point in some cases. For instance, when $n = 2$, ω_n/ω is rational, thus Q is a periodic orbit. When $n = 2$

and $a = \sqrt{2}$, Q is a fixed point

$$Q = -\sqrt{2}e^{i2x+i(2\gamma-\theta)} = qe^{i(2\gamma-2\theta+\pi)}.$$

When $(a^2-2)^2 = \frac{\ell^2}{j^2-\ell^2}(n^2-4)$ for some integers j and ℓ (such that $(a^2-2)^2 < 3$), ω_n/ω is rational, and Q is a periodic orbit.

To see the geometric multiplicity at $\lambda = \nu_n$ for Q, we start with $\phi^\pm$. Under the Darboux transformation in Theorem 7.2, they are transformed into

$$\Phi^\pm = G\phi^\pm = \pm\nu_n c^\mp W(\phi^+, \phi^-) \begin{pmatrix} 1/\phi_1 \\ -1/\phi_2 \end{pmatrix}, \tag{7.11}$$

where $W(\phi^+, \phi^-)$ is the Wronskian of ϕ^+ and ϕ^-.

7.6 Whisker of the Strange $\mathbb{T}^2$

Let $\xi_+ = 1/2$ in (7.6) which corresponds to $n = 1$ in (7.7) and denote the λ by ν_1:

$$\nu_1^2 = \frac{1}{2}(a^2-2) + \frac{i}{2}\sqrt{3-(a^2-2)^2}.$$

Then the eigenfunctions (7.6) have the form

$$\varphi^\pm = \begin{pmatrix} \nu_1 a e^{i\vartheta/2} \\ i\left(\nu_1^2 + 1 \pm \frac{1}{2}\right) e^{-i\vartheta/2} \end{pmatrix} e^{\pm i\frac{1}{2}x + \hat{\Omega}_\pm t} \tag{7.12}$$

where

$$\hat{\Omega}_\pm = \pm\frac{1}{2}\sqrt{3-(a^2-2)^2} \pm i(2-a^2).$$

Let

$$\varphi = c_+\varphi^+ + c_-\varphi^-, \quad c_+/c_- = e^{\rho+i\alpha}. \tag{7.13}$$

Under the Darboux transformation leading to (7.9), φ is transformed into

$$\hat{\varphi} = \begin{pmatrix} \nu_1 \frac{\phi_2}{\phi_1}\varphi_1 - \nu_n\varphi_2 \\ -\nu_n\varphi_1 + \nu_1\frac{\phi_1}{\phi_2}\varphi_2 \end{pmatrix},$$

where $\hat{\varphi}$ solves the Lax Pair (7.2) and (7.3) at (ν_1, Q). Introducing the new notations

$$F = \phi_2/\phi_1, \quad \tilde{F} = \varphi_2/\varphi_1, \quad \hat{F} = \hat{\varphi}_2/\hat{\varphi}_1,$$

we have that F is given in (7.8), $\tilde{F}$ has the expression

$$\tilde{F} = \frac{i}{\nu_1 a}e^{-i\vartheta}\frac{\left(\nu_1^2+\frac{3}{2}\right)\exp(ix+2\hat{\Omega}_+t+\rho+i\alpha) + \nu_1^2 + \frac{1}{2}}{\exp(ix+2\hat{\Omega}_+t+\rho+i\alpha)+1}, \tag{7.14}$$

and $\hat{F}$ is given by

$$\hat{F} = \frac{-\nu_n + \nu_1 \tilde{F}/F}{\nu_1 F - \nu_n \tilde{F}}. \tag{7.15}$$

Applying the Darboux transformation in Theorem 7.3 at (ν_1, Q), one obtains that

$$\tilde{Q} = \frac{\overline{\nu_1} - \nu_1|\hat{F}|^2}{(\overline{\nu_1}|\hat{F}|^2 - \nu_1)^2}\left[(\overline{\nu_1} - \nu_1|\hat{F}|^2)Q + 2i(\nu_1^2 - \overline{\nu_1}^2)\bar{\hat{F}}\right]. \tag{7.16}$$

7.7 Whisker of the Circle

Directly using φ in (7.13) in the Darboux transformation in Theorem 7.3, one generates the whisker of the circle represented by q in (7.5),

$$\hat{Q} = \frac{\overline{\nu_1} - \nu_1|\tilde{F}|^2}{(\overline{\nu_1}|\tilde{F}|^2 - \nu_1)^2}\left[(\overline{\nu_1} - \nu_1|\tilde{F}|^2)q + 2i(\nu_1^2 - \overline{\nu_1}^2)\bar{\tilde{F}}\right]. \tag{7.17}$$

As $t \to +\infty$,

$$\tilde{F} \to ie^{-i\vartheta}\frac{\nu_1^2 + \frac{3}{2}}{\nu_1 a}.$$

As $t \to -\infty$,

$$\tilde{F} \to ie^{-i\vartheta}\frac{\nu_1^2 + \frac{1}{2}}{\nu_1 a}.$$

As $t \to +\infty$,

$$\hat{Q} \to qe^{i\beta_+}, \tag{7.18}$$

where

$$e^{i\beta_+} = \frac{\nu_1^2}{\overline{\nu_1}^2}\frac{\left(\overline{\nu_1}^2 + \frac{3}{2}\right)^2}{\left(\nu_1^2 + \frac{3}{2}\right)^2}\frac{-1 - (\nu_1^2 - \overline{\nu_1}^2)}{-1 + (\nu_1^2 - \overline{\nu_1}^2)}.$$

As $t \to -\infty$,

$$\hat{Q} \to qe^{i\beta_-}, \tag{7.19}$$

where

$$e^{i\beta_-} = \frac{\nu_1^2}{\overline{\nu_1}^2}\frac{\left(\overline{\nu_1}^2 + \frac{1}{2}\right)^2}{\left(\nu_1^2 + \frac{1}{2}\right)^2}\frac{1 - (\nu_1^2 - \overline{\nu_1}^2)}{1 + (\nu_1^2 - \overline{\nu_1}^2)}.$$

Under the Darboux transformation (Theorem 7.3), $\varphi^\pm$ are transformed into

$$\tilde{\varphi}^\pm = G\varphi^\pm = \pm c_\mp \nu_1(\nu_1^2 - \overline{\nu_1}^2)W(\varphi^+, \varphi^-)\begin{pmatrix} \dfrac{-\overline{\varphi_2}}{\overline{\nu_1}|\varphi_2|^2 - \nu_1|\varphi_1|^2} \\ \dfrac{\overline{\varphi_1}}{\overline{\nu_1}|\varphi_1|^2 - \nu_1|\varphi_2|^2} \end{pmatrix}. \tag{7.20}$$

7.8 Diffusion

Here diffusion refers to variations of invariants when DNLS is under perturbations. When the perturbation is Hamiltonian, such a diffusion is often called an Arnold diffusion. Along homoclinics or heteroclinics, such a diffusion is often measured by a Melnikov integral.

First we define the invariants that we are interested in.

Definition 7.4. An important sequence of invariants F_j of the DNLS is defined by

$$F_j(q,\bar{q}) = \Delta(\lambda_j^{(c)}(q,\bar{q}),q,\bar{q}).$$

Lemma 7.5. *If* $\{\lambda_j^{(c)}\}$ *is a simple critical point of* Δ*, then*

$$\frac{\partial F_j}{\partial w} = \left.\frac{\partial \Delta}{\partial w}\right|_{\lambda=\lambda_j^{(c)}}, \qquad w = q,\bar{q}.$$

Proof: We know that

$$\frac{\partial F_j}{\partial w} = \left.\frac{\partial \Delta}{\partial w}\right|_{\lambda=\lambda_j^{(c)}} + \left.\frac{\partial \Delta}{\partial \lambda}\right|_{\lambda=\lambda_j^{(c)}} \frac{\partial \lambda_j^{(c)}}{\partial w}.$$

Since

$$\left.\frac{\partial \Delta}{\partial \lambda}\right|_{\lambda=\lambda_j^{(c)}} = 0,$$

we have

$$\left.\frac{\partial^2 \Delta}{\partial \lambda^2}\right|_{\lambda=\lambda_j^{(c)}} \frac{\partial \lambda_j^{(c)}}{\partial w} + \left.\frac{\partial^2 \Delta}{\partial \lambda \partial w}\right|_{\lambda=\lambda_j^{(c)}} = 0.$$

Since $\lambda_j^{(c)}$ is a simple critical point of Δ,

$$\left.\frac{\partial^2 \Delta}{\partial \lambda^2}\right|_{\lambda=\lambda_j^{(c)}} \neq 0.$$

Thus

$$\frac{\partial \lambda_j^{(c)}}{\partial w} = -\left(\left.\frac{\partial^2 \Delta}{\partial \lambda^2}\right|_{\lambda=\lambda_j^{(c)}}\right)^{-1} \left.\frac{\partial^2 \Delta}{\partial \lambda \partial w}\right|_{\lambda=\lambda_j^{(c)}}.$$

Notice that Δ is an entire function of λ and $w = q,\bar{q}$ [76], then we know that $\frac{\partial \lambda_j^{(c)}}{\partial w}$ is bounded, and

$$\frac{\partial F_j}{\partial w} = \left.\frac{\partial \Delta}{\partial w}\right|_{\lambda=\lambda_j^{(c)}}.$$

□

Theorem 7.6. *As a function of three variables, $\Delta = \Delta(\lambda; q; \bar{q})$ has the partial derivatives given by Bloch functions $\psi^{\pm}$ (i.e., $\psi^{\pm}(x) = e^{\pm\Lambda x}\tilde{\psi}^{\pm}(x)$, where $\tilde{\psi}^{\pm}$ are periodic in x of period 2π, and Λ is a complex constant):*

$$\frac{\partial \Delta}{\partial q} = \frac{\lambda\sqrt{\Delta^2-4}}{W(\psi^+,\psi^-)}\psi_2^+\psi_2^-, \tag{7.21}$$

$$\frac{\partial \Delta}{\partial \bar{q}} = -\frac{\lambda\sqrt{\Delta^2-4}}{W(\psi^+,\psi^-)}\psi_1^+\psi_1^-,$$

$$\frac{\partial \Delta}{\partial \lambda} = \frac{\sqrt{\Delta^2-4}}{W(\psi^+,\psi^-)}\int_0^{2\pi}[q\psi_2^+\psi_2^- - \bar{q}\psi_1^+\psi_1^- - i2\lambda(\psi_1^+\psi_2^- + \psi_2^+\psi_1^-)]dx. \tag{7.22}$$

where $W(\psi^+,\psi^-) = \psi_1^+\psi_2^- - \psi_2^+\psi_1^-$ is the Wronskian.

Proof: Recall that M is the fundamental matrix solution of (7.2), we have the equation for the differential of M

$$\partial_x dM = UdM + dUM, \quad dM(0) = 0.$$

Using the method of variation of parameters, we let

$$dM = MR, \quad R(0) = 0.$$

Thus

$$R(x) = \int_0^x M^{-1}dUMdx,$$

and

$$dM(x) = M(x)\int_0^x M^{-1}dUMdx.$$

Finally

$$\begin{aligned} d\Delta &= \text{ trace } dM(2\pi) \\ &= \text{ trace } \left(M(2\pi)\int_0^{2\pi} M^{-1}dUMdx\right). \end{aligned} \tag{7.23}$$

Let

$$N = (\psi^+\psi^-)$$

where $\psi^{\pm}$ are two linearly independent Bloch functions (For the case that there is only one linearly independent Bloch function, L'Hospital's rule has to be used, for details, see [76]), such that

$$\psi^{\pm} = e^{\pm\Lambda x}\tilde{\psi}^{\pm},$$

where $\tilde{\psi}^{\pm}$ are periodic in x of period 2π and Λ is a complex constant (The existence of such functions is the result of the well known Floquet theorem). Then

$$N(x) = M(x)N(0), \quad M(x) = N(x)[N(0)]^{-1}.$$

Notice that

$$N(2\pi) = N(0)E, \quad \text{where } E = \begin{pmatrix} e^{\Lambda 2\pi} & 0 \\ 0 & e^{-\Lambda 2\pi} \end{pmatrix}.$$

Then

$$M(2\pi) = N(0)E[N(0)]^{-1}.$$

Thus

$$\Delta = \text{ trace } M(2\pi) = \text{ trace } E = e^{\Lambda 2\pi} + e^{-\Lambda 2\pi},$$

and

$$e^{\pm\Lambda 2\pi} = \frac{1}{2}(\Delta \pm \sqrt{\Delta^2 - 4}).$$

In terms of N, $d\Delta$ as given in (7.23) takes the form

$$\begin{aligned} d\Delta = &\text{ trace } \left\{ N(0)E[N(0)]^{-1} \int_0^{2\pi} N(0)[N(x)]^{-1} dU(x)N(x)[N(0)]^{-1} dx \right\} \\ = &\text{ trace } \left\{ E \int_0^{2\pi} [N(x)]^{-1} dU(x)N(x)dx \right\}, \end{aligned}$$

from which one obtains the partial derivatives of Δ as stated in the theorem. □

Consider a perturbed derivative nonlinear Schrödinger equation,

$$iq_t = q_{xx} - i\left(|q|^2 q\right)_x + i\epsilon f, \tag{7.24}$$

where ϵ is the perturbation parameter, and f may depend on q and its spatial derivatives, x and t. Then we have the diffusion formula,

$$\begin{aligned} \frac{dF_j}{dt} &= \int_0^{2\pi} \left(\frac{\partial F_j}{\partial q} q_t + \frac{\partial F_j}{\partial \bar{q}} \bar{q}_t \right) dx \\ &= \epsilon \int_0^{2\pi} \left(\frac{\partial F_j}{\partial q} f + \frac{\partial F_j}{\partial \bar{q}} \bar{f} \right) dx \\ &= \epsilon \frac{\lambda_j^{(c)} \sqrt{\Delta^2 - 4}}{W(\psi^+, \psi^-)} \int_0^{2\pi} \left[\left(\psi_2^+ \psi_2^-\right) f - \left(\psi_1^+ \psi_1^-\right) \bar{f} \right] dx. \end{aligned} \tag{7.25}$$

Here F_j Poisson commutes with the Hamiltonian of DNLS. For any solution $q(t, x)$ of (7.24), at any fixed time t, $\frac{\partial F_j}{\partial q}$ evaluated at $q(t, x)$ is given by (7.21); similarly for $\frac{\partial F_j}{\partial \bar{q}}$. For specific examples, the leading order term of (7.25) can be calculated.

7.9 Diffusion Along the Strange $\mathbb{T}^2$

Here we are interested in the diffusion of F_n with $\lambda_n^{(c)} = \nu_n$ $(n \geq 2)$. First we need to evaluate the gradient of F_n as given in (7.21) and (7.22) along the strange $\mathbb{T}^2$ (7.9). In the neighborhood of ν_n $(n \geq 1)$, to the leading order

$$\Delta^2 - 4 = (\Delta + 2)(\Delta - 2) = \Delta(\nu_n)\Delta''(\nu_n)(\lambda - \nu_n)^2, \tag{7.26}$$

where

$$\Delta(\nu_n) = 2(-1)^n, \quad \Delta''(\nu_n) = (-1)^n 32\pi^2 \nu_n^2 n^{-2}[2(\nu_n^2+1) - a^2]^2.$$

Under the Darboux transformation in Theorem 7.2,

$$W(\Psi^+, \Psi^-) = (\lambda^2 - \nu_n^2) W(\psi^+, \psi^-),$$

where as $\lambda \to \nu_n$, $\psi^\pm \to \phi^\pm$ and $\Psi^\pm \to \Phi^\pm$ as given in (7.11). By L'Hospital's rule,

$$\lim_{\lambda\to\nu_n} \frac{\sqrt{\Delta^2(\lambda)-4}}{W(\Psi^+,\Psi^-)} = \frac{\sqrt{\Delta^2(\nu_n)-4}}{W(\Phi^+,\Phi^-)} = \frac{\sqrt{\Delta(\nu_n)\Delta''(\nu_n)}}{2\nu_n W(\phi^+,\phi^-)},$$

Thus

$$\left.\frac{\partial F_n}{\partial q}\right|_{q=Q} = \frac{\sqrt{\Delta(\nu_n)\Delta''(\nu_n)}}{2W(\phi^+,\phi^-)} \Phi_2^+ \Phi_2^-,$$

$$\left.\frac{\partial F_n}{\partial \bar{q}}\right|_{q=Q} = -\frac{\sqrt{\Delta(\nu_n)\Delta''(\nu_n)}}{2W(\phi^+,\phi^-)} \Phi_1^+ \Phi_1^-,$$

where Q is given in (7.9) and $\Phi^\pm$ is given in (7.11). Then, to the leading order in ϵ,

$$\frac{dF_n}{dt} = \epsilon \frac{\sqrt{\Delta(\nu_n)\Delta''(\nu_n)}}{2W(\phi^+,\phi^-)} \int_0^{2\pi} \left[\left(\Phi_2^+\Phi_2^-\right) f(Q) - \left(\Phi_1^+\Phi_1^-\right) \overline{f(Q)} \right] dx.$$

For a finite T, the leading order diffusion of F_n is given by

$$\int_0^T \frac{dF_n}{dt} dt = \epsilon \frac{\sqrt{\Delta(\nu_n)\Delta''(\nu_n)}}{2W(\phi^+,\phi^-)} \int_0^T \int_0^{2\pi} \left[\left(\Phi_2^+\Phi_2^-\right) f(Q) - \left(\Phi_1^+\Phi_1^-\right) \overline{f(Q)} \right] dx dt.$$

7.10 Diffusion Along the Whisker of the Circle

Notice that here under the Darboux transformation in Theorem 7.3,

$$\det G = (\lambda^2 - \nu_1^2)(\lambda^2 - \overline{\nu_1}^2).$$

Thus by the L'Hospital's rule as above,

$$\frac{\sqrt{\Delta^2(\nu_1)-4}}{W(\tilde{\varphi}^+,\tilde{\varphi}^-)} = \frac{\sqrt{\Delta(\nu_1)\Delta''(\nu_1)}}{2\nu_1(\nu_1^2 - \overline{\nu_1}^2) W(\varphi^+,\varphi^-)},$$

where $\tilde{\varphi}^\pm$ is given in (7.20). Thus

$$\left.\frac{\partial F_1}{\partial q}\right|_{q=\hat{Q}} = \frac{\sqrt{\Delta(\nu_1)\Delta''(\nu_1)}}{2(\nu_1^2 - \overline{\nu_1}^2) W(\varphi^+,\varphi^-)} \tilde{\varphi}_2^+ \tilde{\varphi}_2^-,$$

$$\left.\frac{\partial F_1}{\partial \bar{q}}\right|_{q=\hat{Q}} = -\frac{\sqrt{\Delta(\nu_1)\Delta''(\nu_1)}}{2(\nu_1^2 - \overline{\nu_1}^2) W(\varphi^+,\varphi^-)} \tilde{\varphi}_1^+ \tilde{\varphi}_1^-,$$

where $\hat{Q}$ is given in (7.17). Then, to the leading order in ϵ,

$$\frac{dF_1}{dt} = \epsilon \frac{\sqrt{\Delta(\nu_1)\Delta''(\nu_1)}}{2(\nu_1^2 - \overline{\nu_1}^2)W(\varphi^+, \varphi^-)} \int_0^{2\pi} \left[\left(\tilde{\varphi}_2^+ \tilde{\varphi}_2^-\right) f(\hat{Q}) - \left(\tilde{\varphi}_1^+ \tilde{\varphi}_1^-\right) \overline{f(\hat{Q})} \right] dx.$$

From their expressions (7.20), as $|t| \to \infty$, $\tilde{\varphi}^\pm \to 0$. Also notice the asymptotics of $\hat{Q}$ in (7.18) and (7.19). In this case, often the infinite time leading order diffusion of F_1 can be given by the Melnikov integral

$$\int_{-\infty}^{\infty} \frac{dF_1}{dt} dt = \epsilon \frac{\sqrt{\Delta(\nu_1)\Delta''(\nu_1)}}{2(\nu_1^2 - \overline{\nu_1}^2)W(\varphi^+, \varphi^-)} \int_{-\infty}^{\infty} \int_0^{2\pi} \left[\left(\tilde{\varphi}_2^+ \tilde{\varphi}_2^-\right) f(\hat{Q}) \right.$$
$$\left. - \left(\tilde{\varphi}_1^+ \tilde{\varphi}_1^-\right) \overline{f(\hat{Q})} \right] dx dt.$$

Chapter 8

Discrete Nonlinear Schrödinger Equation

8.1 Background

The discrete nonlinear Schrödinger equation models the dynamics of a lattice of anharmonic oscillators. Its continuous version is the well-known nonlinear Schrödinger equation. It is actually a finite difference discretization of the nonlinear Schrödinger equation. Take the integrable nonlinear Schrödinger equation as the example

$$iq_t = q_{xx} + 2|q|^2 q.$$

There are two natural discretizations, the diagonal one

$$iq_{n,t} = \frac{1}{h^2}(q_{n+1} - 2q_n + q_{n-1}) + 2|q_n|^2 q_n,$$

and the central one

$$iq_{n,t} = \frac{1}{h^2}(q_{n+1} - 2q_n + q_{n-1}) + |q_n|^2(q_{n+1} + q_{n-1});$$

where h is the mesh length. It turns out that the diagonal one is not integrable, while the central one is [1]. In fact the dynamics of the diagonal one is chaotic [43]. This chapter will focus on the central one. For more details on this chapter, we refer the readers to [60, 65, 77].

8.2 Hamiltonian Structure

We consider the discrete nonlinear Schrödinger equation (DNLS)

$$i\dot{q}_n = \frac{1}{h^2}(q_{n+1} - 2q_n + q_{n-1}) + |q_n|^2(q_{n+1} + q_{n-1}) - 2\omega^2 q_n, \tag{8.1}$$

where $i = \sqrt{-1}$, q_n's are complex variables, $n \in \mathbb{Z}$, and ω is a positive parameter; under periodic boundary condition and even constraint,

$$q_{n+N} = q_n, \quad q_{-n} = q_n.$$

The DNLS can be re-written in the Hamiltonian form

$$i\dot{q}_n = \rho_n \partial H / \partial \bar{q}_n, \tag{8.2}$$

where

$$H = \frac{1}{h^2} \sum_{n=0}^{N-1} \left[\bar{q}_n (q_{n+1} + q_{n-1}) - \frac{2}{h^2}(1 + \omega^2 h^2) \ln \rho_n \right],$$

and $\rho_n = 1 + h^2 |q_n|^2$. Notice that $\sum_{n=0}^{N-1} [\bar{q}_n (q_{n+1} + q_{n-1})]$ itself is also a constant of motion. This invariant, together with H, implies that $\sum_{n=0}^{N-1} \ln \rho_n$ is a constant of motion too. Therefore,

$$D^2 \equiv \prod_{n=0}^{N-1} \rho_n \tag{8.3}$$

is a constant of motion.

8.3 Lax Pair and Floquet Theory

The DNLS (8.1) has the discretized Lax pair [1]:

$$\varphi_{n+1} = L_n^{(z)} \varphi_n, \tag{8.4}$$

$$\dot{\varphi}_n = B_n^{(z)} \varphi_n, \tag{8.5}$$

where

$$L_n^{(z)} = \begin{pmatrix} z & ihq_n \\ ih\bar{q}_n & 1/z \end{pmatrix}, \quad B_n^{(z)} = \frac{i}{h^2} \begin{pmatrix} B_{11} & B_{12} \\ B_{21} & B_{22} \end{pmatrix},$$

$$\begin{aligned} B_{11} &= 1 - z^2 + 2i\lambda h - h^2 q_n \bar{q}_{n-1} + \omega^2 h^2, \\ B_{12} &= -izhq_n + (1/z) ihq_{n-1}, \\ B_{21} &= -izh\bar{q}_{n-1} + (1/z) ih\bar{q}_n, \\ B_{22} &= 1/z^2 - 1 + 2i\lambda h + h^2 \bar{q}_n q_{n-1} - \omega^2 h^2, \end{aligned}$$

and $z = \exp(i\lambda h)$. Compatibility of the over determined system (8.4) and (8.5) gives the "Lax representation"

$$\dot{L}_n = B_{n+1} L_n - L_n B_n$$

of the discrete DNLS (8.1). Focusing attention upon the discrete spatial flow (8.4), we let $Y^{(1)}, Y^{(2)}$ be the fundamental solutions of the ODE (8.4), i.e., solutions with the initial conditions:

$$Y_0^{(1)} = \begin{pmatrix} 1 \\ 0 \end{pmatrix}, \quad Y_0^{(2)} = \begin{pmatrix} 0 \\ 1 \end{pmatrix}.$$

The Floquet discriminant is defined by

$$\Delta(z;\vec{q}) \equiv \mathrm{tr}(M(N;z;\vec{q})), \tag{8.6}$$

where $M(n;z;\vec{q}) = \text{columns}\{Y_n^{(1)}, Y_n^{(2)}\}$ is the fundamental matrix of (8.4).

Remark 8.1. $\Delta(z;\vec{q})$ is a constant of motion for the integrable system (8.1) for any $z \in \mathbb{C}$. Since $\Delta(z;\vec{q})$ is a meromorphic function in z of degree $(+N,-N)$, the Floquet discriminant $\Delta(z;\vec{q})$ acts as a generating function for $(M+1)$ functionally independent constants of motion, and is the key to the complete integrability of the system (8.1), where $M = N/2$ (N even), $M = (N-1)/2$ (N odd).

The Floquet theory here is not standard as can be seen from the Wronskian relation:

$$W_N(\psi^+,\psi^-) = D^2\ W_0(\psi^+,\psi^-),$$

where D is defined in (8.3),

$$W_n(\psi^+,\psi^-) \equiv \psi_n^{(+,1)}\psi_n^{(-,2)} - \psi_n^{(+,2)}\psi_n^{(-,1)},$$

ψ^+ and ψ^- are any two solutions to the linear system (8.4). In fact, $W_{n+1}(\psi^+,\psi^-) = \rho_n W_n(\psi^+,\psi^-)$. Due to this nonstandardness, modifications of the usual definitions of spectral quantities [60] are required. The Floquet spectrum is defined as the closure of the complex λ for which there exists a bounded eigenfunction to the ODE (8.4). In terms of the Floquet discriminant Δ, this is given by

$$\sigma(L) = \{z \in C | -2D \leq \Delta(z;\vec{q}) \leq 2D\}.$$

Periodic and antiperiodic points z^s are defined by

$$\Delta(z^s;\vec{q}) = \pm 2D.$$

A critical point z^c is defined by the condition

$$\left.\frac{d\Delta}{dz}\right|_{(z^c;\vec{q})} = 0.$$

A multiple point z^m is a critical point which is also a periodic or antiperiodic point. The *algebraic multiplicity* of z^m is defined as the order of the zero of $\Delta(z) \pm 2D$. Usually it is 2, but it can exceed 2; when it does equal 2, we call the multiple point a *double point*, and denote it by z^d. The *geometric multiplicity* of z^m is defined as the dimension of the periodic (or antiperiodic) eigenspace of (8.4) at z^m, and is either 1 or 2.

The normalized Floquet discriminant $\tilde{\Delta}$ is defined as

$$\tilde{\Delta} = \Delta / D.$$

Remark 8.2 (Continuum Limit)**.** In the continuum limit (i.e., $h \to 0$), the Hamiltonian has a limit in the manner: $H/h \to H_c$, where H_c is the Hamiltonian for NLS PDE studied in Chapter 3, $H_c = i\int_0^1 [q_x \bar{q}_x + 2\omega^2|q|^2 - |q|^4]dx$. The Lax pair (8.4) and (8.5) also tends to the corresponding Lax pair for NLS PDE with spectral parameter λ ($z = e^{i\lambda h}$) [76]. If $Q \equiv \max_n\{|q_n|\}$ is finite, then $\rho_n \to 1$ as $h \to 0$. Therefore, $D^2 \equiv \{\prod_{n=0}^{N-1} \rho_n\} \to 1$ as $h \to 0$. The nonstandard Floquet theory for the spatial part of the Lax pair (8.4) and (8.5) becomes the standard Floquet theory in the continuum limit.

8.4 Examples of Floquet Spectra

Consider the uniform solution: $q_n = q_c$, $\forall n$

$$q_c(t) = a\exp\{-i[2(a^2 - \omega^2)t - \gamma]\}.$$

The corresponding Bloch functions of the Lax pair are given by:

$$\psi_n^+ = (\sqrt{\rho}e^{i\beta})^n e^{\Omega_+ t}\begin{pmatrix}(1/z - \sqrt{\rho}e^{i\beta})\exp\{-i[(a^2-\omega^2)t - \gamma/2]\}\\ -iha\exp\{i[(a^2-\omega^2)t-\gamma/2]\}\end{pmatrix},$$

$$\psi_n^- = (\sqrt{\rho}e^{-i\beta})^n e^{\Omega_- t}\begin{pmatrix}-iha\exp\{-i[(a^2-\omega^2)t - \gamma/2]\}\\ (z - \sqrt{\rho}e^{-i\beta})\exp\{i[(a^2-\omega^2)t-\gamma/2]\}\end{pmatrix},$$

where

$$\begin{aligned} z &= \sqrt{\rho}\cos\beta + \sqrt{\rho\cos^2\beta - 1}, \quad \rho = 1 + h^2a^2, \\ \Omega_+ &= i/h^2[(1/z - z)\sqrt{\rho}e^{i\beta} + i2\lambda h], \\ \Omega_- &= i/h^2[(1/z - z)\sqrt{\rho}e^{-i\beta} + i2\lambda h]. \end{aligned}$$

The Floquet discriminant is given by:

$$\Delta = 2D\cos(N\beta),$$

where $D = \rho^{N/2}$. Thus the Floquet spectra are given by:

$$-1 \le \cos(N\beta) \le 1.$$

Periodic and antiperiodic points are given by:

$$z_m^{(s)} = \sqrt{\rho}\cos\frac{m}{N}\pi + \sqrt{\rho\cos^2\frac{m}{N}\pi - 1},$$

where $z_m^{(s)}$ is a periodic point when m is even, and $z_m^{(s)}$ is an antiperiodic point when m is odd. The following facts are obvious,

- If N is odd, all the periodic and antiperiodic points are on the real axis for sufficiently large $|q_c|$.
- If N is even, except the two points $z = \pm i$, all other periodic and antiperiodic points are on the real axis for sufficiently large $|q_c|$.

Derivatives of the Floquet discriminant Δ with respect to z are given by:

$$\begin{aligned} d\Delta/dz &= 2ND\sin(N\beta)(z\sqrt{\rho}\sin\beta)^{-1}\sqrt{\rho\cos^2\beta - 1}, \\ d^2\Delta/dz^2 &= -2ND(\rho z^2\sin^3\beta)^{-1}[N\cos(N\beta)\sin\beta\ (\rho\cos^2\beta - 1) \\ &\quad +(1-\rho)\cos\beta\sin(N\beta) + \sqrt{\rho}\sin(N\beta)\sin^2\beta\sqrt{\rho\cos^2\beta - 1}]. \end{aligned}$$

The critical points are given by:

$$\beta = \frac{m}{N}\pi \quad (\beta \neq 0, \pi), \quad \text{or} \ \cos^2\beta = 1/\rho.$$

There can be multiple points with algebraic multiplicity greater than 2. For example, when two symmetric double points on the circle collide at the intersection points $z = \pm 1$, we have multiple points of algebraic multiplicity 4. In such case, $\rho\cos^2\beta - 1 = 0$ and $\sin(N\beta) = 0$; then $d^2\Delta/dz^2 = 0$, at $z = \pm 1$.

8.5 Melnikov Vectors

Definition 8.3. The sequence of invariants $\tilde{F}_j$ is defined as:

$$\tilde{F}_j(\vec{q}) = \tilde{\Delta}(z_j^{(c)}(\vec{q}); \vec{q}). \tag{8.7}$$

These invariants $\tilde{F}_j$'s are perfect candidate for building Melnikov functions. The Melnikov vectors are given by the gradients of these invariants.

Lemma 8.4. *Let $z_j^{(c)}(\vec{q})$ be a simple critical point; then*

$$\frac{\delta \tilde{F}_j}{\delta \vec{q}_n}(\vec{q}) = \frac{\delta \tilde{\Delta}}{\delta \vec{q}_n}(z_j^{(c)}(\vec{q}); \vec{q}). \tag{8.8}$$

$$\frac{\delta \tilde{\Delta}}{\delta \vec{q}_n}(z; \vec{q}) = \frac{ih(\zeta - \zeta^{-1})}{2W_{n+1}} \begin{pmatrix} \psi_{n+1}^{(+,2)}\psi_n^{(-,2)} + \psi_n^{(+,2)}\psi_{n+1}^{(-,2)} \\ \psi_{n+1}^{(+,1)}\psi_n^{(-,1)} + \psi_n^{(+,1)}\psi_{n+1}^{(-,1)} \end{pmatrix}, \tag{8.9}$$

where $\psi_n^{\pm} = (\psi_n^{(\pm,1)}, \psi_n^{(\pm,2)})^T$ are two Bloch functions of the Lax pair (8.4) and (8.5), such that

$$\psi_n^{\pm} = D^{n/N}\zeta^{\pm n/N}\tilde{\psi}_n^{\pm},$$

where $\tilde{\psi}_n^{\pm}$ are periodic in n with period N, $W_n = \det\,(\psi_n^+, \psi_n^-)$.

Proof: By the definition of critical points,

$$\tilde{\Delta}'(z_j^{(c)}(\vec{q});\vec{q}) = 0.$$

Differentiating this equation, we have

$$\frac{\delta z_j^{(c)}}{\delta \vec{q}_n} = -\frac{1}{\tilde{\Delta}''}\frac{\delta \tilde{\Delta}'}{\delta \vec{q}_n}.$$

Since $z_j^{(c)}(\vec{q})$ is a simple critical point, $z_j^{(c)}$ is a differentiable function. Thus

$$\begin{aligned}\frac{\delta \tilde{F}_j}{\delta \vec{q}_n} &= \left.\frac{\delta \tilde{\Delta}}{\delta \vec{q}_n}\right|_{z=z_j^{(c)}} + \left.\frac{\partial \tilde{\Delta}}{\partial z}\right|_{z=z_j^{(c)}} \frac{\delta z_j^{(c)}}{\delta \vec{q}_n} \\ &= \left.\frac{\delta \tilde{\Delta}}{\delta \vec{q}_n}\right|_{z=z_j^{(c)}}.\end{aligned}$$

This proves equation (8.8). Next we derive the formula (8.9). Let M_n be the fundamental matrix to the system (8.4), i.e., the matrix solution to (8.4) with initial condition M_0 being a 2×2 identity matrix. Variation of $\vec{q}$ leads to the variational equation for the variation of M_n at fixed z,

$$\begin{aligned}\delta M_{n+1} &= \begin{pmatrix} z & ihq_n \\ ih\bar{q}_n & 1/z \end{pmatrix} \delta M_n + \begin{pmatrix} 0 & ih\,\delta q_n \\ ih\,\overline{\delta q_n} & 0 \end{pmatrix} M_n, \\ \delta M_0 &= 0.\end{aligned}$$

Let $\delta M_n = M_n A_n$, where A_n is a 2×2 matrix to be determined, we have

$$\begin{aligned} A_{n+1} - A_n &= M_{n+1}^{-1}\delta U_n M_n, \\ A_0 &= 0, \end{aligned} \tag{8.10}$$

where

$$\delta U_n = \begin{pmatrix} 0 & ih\delta q_n \\ ih\overline{\delta q_n} & 0 \end{pmatrix}.$$

Solving the system (8.10), we have

$$\begin{aligned}\delta M_n &= M_n \left(\sum_{j=1}^{n} M_j^{-1}\delta U_{j-1} M_{j-1} \right), \\ \delta M_0 &= 0.\end{aligned}$$

Then,

$$\delta \Delta(z,\vec{q}) = \text{trace}\left(M_N \left(\sum_{j=1}^{N} M_j^{-1}\delta U_{j-1} M_{j-1} \right)\right).$$

Thus,

$$\frac{\delta\Delta}{\delta q_n} = ih \text{ trace}\left(M_{n+1}^{-1} \begin{pmatrix} 0 & 1 \\ 0 & 0 \end{pmatrix} M_n M_N \right), \tag{8.11}$$

$$\frac{\delta\Delta}{\delta r_n} = -ih \text{ trace}\left(M_{n+1}^{-1} \begin{pmatrix} 0 & 0 \\ 1 & 0 \end{pmatrix} M_n M_N \right), \tag{8.12}$$

where $r_n = -\bar{q}_n$. Let ψ^+ and ψ^- be two Bloch functions for the discrete Lax pairs (8.4) and (8.5),

$$\psi_n^{\pm} = D^{n/N}\zeta^{\pm n/N}\tilde{\psi}_n^{\pm}, \tag{8.13}$$

where $\tilde{\psi}_n^{\pm}$ are periodic in n with period N. Let B_n be the 2×2 matrix with ψ_n^+ and ψ_n^- as the column vectors,

$$B_n = (\psi_n^+, \psi_n^-). \tag{8.14}$$

Then

$$M_n = B_n B_0^{-1}, \quad M_N = B_0 \begin{pmatrix} D\zeta & 0 \\ 0 & D\zeta^{-1} \end{pmatrix} B_0^{-1}. \tag{8.15}$$

Substitute the representations (8.14) and (8.15) into (8.11) and (8.12), we have

$$\frac{\delta\Delta}{\delta\vec{q}_n} = \tilde{\Delta}\frac{\delta D}{\delta\vec{q}_n} + \frac{iDh(\zeta - \zeta^{-1})}{2W_{n+1}} \begin{pmatrix} \psi_{n+1}^{(+,2)}\psi_n^{(-,2)} + \psi_n^{(+,2)}\psi_{n+1}^{(-,2)} \\ \psi_{n+1}^{(+,1)}\psi_n^{(-,1)} + \psi_n^{(+,1)}\psi_{n+1}^{(-,1)} \end{pmatrix},$$

where

$$W_n = \det B_n, \quad \psi_n^{\pm} = (\psi_n^{(\pm,1)}, \psi_n^{(\pm,2)})^T.$$

Thus,

$$\frac{\delta\tilde{\Delta}}{\delta\vec{q}_n} = \frac{ih(\zeta - \zeta^{-1})}{2W_{n+1}} \begin{pmatrix} \psi_{n+1}^{(+,2)}\psi_n^{(-,2)} + \psi_n^{(+,2)}\psi_{n+1}^{(-,2)} \\ \psi_{n+1}^{(+,1)}\psi_n^{(-,1)} + \psi_n^{(+,1)}\psi_{n+1}^{(-,1)} \end{pmatrix},$$

which is the formula (8.9). The lemma is proved. □

8.6 Darboux Transformations

The hyperbolic structure and homoclinic orbits for (8.1) are constructed through the Darboux transformations, which were built in [60]. First, we present the Darboux transformations. Then, we show how to construct homoclinic orbits.

Fix a solution $q_n(t)$ of the system (8.1), for which the linear operator L_n has a double point z^d of geometric multiplicity 2, which is not on the unit circle. We denote two linearly independent solutions (Bloch functions) of the discrete Lax pair (8.4) and (8.5) at $z = z^d$ by (ϕ_n^+, ϕ_n^-). Thus, a general solution of the discrete Lax pair (8.4) and (8.5) at $(q_n(t), z^d)$ is given by

$$\phi_n(t; z^d, c^+, c^-) = c^+\phi_n^+ + c^-\phi_n^-,$$

where c^+ and c^- are complex parameters. We use ϕ_n to define a transformation matrix Γ_n by

$$\Gamma_n = \begin{pmatrix} z + (1/z)a_n & b_n \\ c_n & -1/z + zd_n \end{pmatrix},$$

where,

$$\begin{aligned} a_n &= \frac{z^d}{(\bar{z}^d)^2\Delta_n}(|\phi_{n2}|^2 + |z^d|^2|\phi_{n1}|^2), \\ d_n &= -\frac{1}{z^d\Delta_n}(|\phi_{n2}|^2 + |z^d|^2|\phi_{n1}|^2), \\ b_n &= \frac{|z^d|^4 - 1}{(\bar{z}^d)^2\Delta_n}\phi_{n1}\bar{\phi}_{n2}, \\ c_n &= \frac{|z^d|^4 - 1}{z^d\bar{z}^d\Delta_n}\bar{\phi}_{n1}\phi_{n2}, \\ \Delta_n &= -\frac{1}{\bar{z}^d}(|\phi_{n1}|^2 + |z^d|^2|\phi_{n2}|^2). \end{aligned}$$

From these formulae, we see that

$$\bar{a}_n = -d_n, \quad \bar{b}_n = c_n.$$

Then we define Q_n and Ψ_n by

$$Q_n \equiv \frac{i}{h}b_{n+1} - a_{n+1}q_n \tag{8.16}$$

and

$$\Psi_n(t; z) \equiv \Gamma_n(z; z^d; \phi_n)\psi_n(t; z) \tag{8.17}$$

where ψ_n solves the discrete Lax pair (8.4) and (8.5) at $(q_n(t), z)$. Formulas (8.16) and (8.17) are the Bäcklund-Darboux transformations for the potential and eigenfunctions, respectively. We have the following theorem [60].

Theorem 8.5 (Darboux Transformations). *Let $q_n(t)$ denote a solution of the system (8.1), for which the linear operator L_n has a double point z^d of geometric multiplicity 2, which is not on the unit circle and which is associated with an instability. We denote two linearly independent solutions (Bloch functions) of the discrete Lax pair (8.4) and (8.5) at (q_n, z^d) by (ϕ_n^+, ϕ_n^-). We define $Q_n(t)$ and $\Psi_n(t; z)$ by (8.16) and (8.17). Then*

(1) *$Q_n(t)$ is also a solution of the system (8.1). (The evenness of Q_n can be obtained by choosing the complex Bäcklund parameter c^+/c^- to lie on a certain curve, as shown in the example below.)*

(2) $\Psi_n(t;z)$ *solves the discrete Lax pair* (8.4) *and* (8.5) *at* $(Q_n(t), z)$.
(3) $\Delta(z;Q_n) = \Delta(z;q_n)$, *for all* $z \in C$.
(4) $Q_n(t)$ *is homoclinic to* $q_n(t)$ *in the sense that* $Q_n(t) \to e^{i\theta_\pm}\, q_n(t)$, *exponentially as* $\exp(-\sigma|t|)$ *as* $t \to \pm\infty$. *Here* $\theta_\pm$ *are the phase shifts,* σ *is a nonvanishing growth rate associated to the double point* z^d, *and explicit formulas can be developed for this growth rate and for the phase shifts* $\theta_\pm$.

This theorem is quite general, constructing homoclinic solutions from a wide class of starting solutions $q_n(t)$. Its proof is by direct verification [60].

8.7 Homoclinic Orbits and Melnikov Vectors

Through the Darboux transformations,homoclinic orbits Q_n can be generated via the formula (8.16)

$$\begin{aligned} Q_n &= ih^{-1}b_{n+1} - a_{n+1}q_n \\ &= \left(\overline{z^{(d)}}(|\phi_{n+1}^{(1)}|^2 + |z^{(d)}|^2|\phi_{n+1}^{(2)}|^2)\right)^{-1} \\ &\quad \times \left(ih^{-1}(1-|z^{(d)}|^4)\phi_{n+1}^{(1)}\overline{\phi_{n+1}^{(2)}} + z^{(d)}q_n(|\phi_{n+1}^{(2)}|^2 + |z^{(d)}|^2|\phi_{n+1}^{(1)}|^2)\right). \end{aligned}$$

The Melnikov vector field located on this homoclinic orbit is given by

$$\frac{\delta\tilde{\Delta}(z^{(d)};\vec{Q})}{\delta\vec{Q}_n} = K\frac{W_n}{E_nA_{n+1}}\begin{pmatrix} [z^{(d)}]^{-2}\,\overline{\phi_n^{(1)}}\,\overline{\phi_{n+1}^{(1)}} \\ \overline{[z^{(d)}]^{-2}}\,\overline{\phi_n^{(2)}}\,\overline{\phi_{n+1}^{(2)}} \end{pmatrix}, \tag{8.18}$$

where

$$\begin{aligned} \phi_n = (\phi_n^{(1)},\phi_n^{(2)})^T &= c_+\psi_n^+ + c_-\psi_n^-, \\ W_n &= \begin{vmatrix}\psi_n^+ & \psi_n^-\end{vmatrix}, \\ E_n &= |\phi_n^{(1)}|^2 + |z^{(d)}|^2|\phi_n^{(2)}|^2, \\ A_n &= |\phi_n^{(2)}|^2 + |z^{(d)}|^2|\phi_n^{(1)}|^2, \\ K = -\frac{ihc_+c_-}{2}&|z^{(d)}|^4(|z^{(d)}|^4-1)[\overline{z^{(d)}}]^{-1}\sqrt{\tilde{\Delta}(z^{(d)};\vec{q})\tilde{\Delta}''(z^{(d)};\vec{q})}, \end{aligned}$$

where

$$\tilde{\Delta}''(z^{(d)};\vec{q}) = \frac{\partial^2\tilde{\Delta}(z^{(d)};\vec{q})}{\partial z^2}.$$

Next we shall study a specific example. we start with the uniform solution of (8.1)

$$q_n = q_c,\ \forall n; \quad q_c = a\exp\{-i[2(a^2-\omega^2)t-\gamma]\}. \tag{8.19}$$

We choose the amplitude a in the range

$$N \tan \frac{\pi}{N} < a < N \tan \frac{2\pi}{N}, \quad \text{when } N > 3,$$
$$3 \tan \frac{\pi}{3} < a < \infty, \quad \text{when } N = 3 \;;$$

so that there is only one set of quadruplets of double points which are not on the unit circle, and denote one of them by $z = z_1^{(d)} = z_1^{(c)}$ which corresponds to $\beta = \pi/N$. The homoclinic orbit Q_n is given by

$$Q_n = q_c(\hat{E}_{n+1})^{-1}(\hat{A}_{n+1} - 2\cos\beta\sqrt{\rho\cos^2\beta - 1}\hat{B}_{n+1}), \tag{8.20}$$

and the Melnikov vectors evaluated on this homoclinic orbit are given by

$$\frac{\delta \tilde{F}_1}{\delta \vec{Q}_n} = \hat{K}(\hat{E}_n \hat{A}_{n+1})^{-1} \text{ sech } (2\mu t + 2p) \begin{pmatrix} \hat{X}_n^{(1)} \\ -\hat{X}_n^{(2)} \end{pmatrix}, \tag{8.21}$$

where

$$\begin{aligned}
\hat{E}_n &= ha\cos\beta + \sqrt{\rho\cos^2\beta - 1} \text{ sech } (2\mu t + 2p)\cos[(2n-1)\beta + \vartheta], \\
\hat{A}_{n+1} &= ha\cos\beta + \sqrt{\rho\cos^2\beta - 1} \text{ sech } (2\mu t + 2p)\cos[(2n+3)\beta + \vartheta], \\
\hat{B}_{n+1} &= \cos\varphi + i\sin\varphi\tanh\,(2\mu t + 2p) + \text{ sech } (2\mu t + 2p)\cos[2(n+1)\beta + \vartheta], \\
\hat{X}_n^{(1)} &= [\cos\beta \text{ sech } (2\mu t + 2p) + \cos[(2n+1)\beta + \vartheta + \varphi] \\
&\quad -i\tanh\,(2\mu t + 2p)\sin[(2n+1)\beta + \vartheta + \varphi]]e^{i2\theta(t)}, \\
\hat{X}_n^{(2)} &= [\cos\beta \text{ sech } (2\mu t + 2p) + \cos[(2n+1)\beta + \vartheta - \varphi] \\
&\quad -i\tanh\,(2\mu t + 2p)\sin[(2n+1)\beta + \vartheta - \varphi]]e^{-i2\theta(t)}, \\
\hat{K} &= -2Nh^2a(1-z^4)(8\rho^{3/2}z^2)^{-1}\sqrt{\rho\cos^2\beta - 1}, \\
\beta &= \pi/N, \quad \rho = 1 + h^2a^2, \quad \mu = 2h^{-2}\sqrt{\rho}\sin\beta\sqrt{\rho\cos^2\beta - 1}, \\
h &= 1/N, \quad c_+/c_- = ie^{2p}e^{i\vartheta}, \quad \vartheta \in [0, 2\pi], \quad p \in (-\infty, \infty), \\
z &= \sqrt{\rho}\cos\beta + \sqrt{\rho\cos^2\beta - 1}, \quad \theta(t) = (a^2 - \omega^2)t - \gamma/2, \\
&\quad \sqrt{\rho\cos^2\beta - 1} + i\sqrt{\rho}\sin\beta = hae^{i\varphi},
\end{aligned}$$

where $\varphi = \sin^{-1}[\sqrt{\rho}(ha)^{-1}\sin\beta]$, $\varphi \in (0, \pi/2)$.

Next we study the "evenness" condition: $Q_{-n} = Q_n$. It turns out that the choices $\vartheta = -\beta$, $-\beta + \pi$ in the formula of Q_n lead to the evenness of Q_n in n. In terms of figure eight structure of Q_n, $\vartheta = -\beta$ corresponds to one ear of the figure eight, and $\vartheta = -\beta + \pi$ corresponds to the other ear. The even formula for Q_n is given by,

$$Q_n = q_c(\Gamma/\Lambda_n - 1), \tag{8.22}$$

where

$$\Gamma = 1 - \cos 2\varphi - i \sin 2\varphi \tanh(2\mu t + 2p), \tag{8.23}$$
$$\Lambda_n = 1 \pm \cos\varphi(\cos\beta)^{-1} \operatorname{sech}(2\mu t + 2p)\cos(2n\beta), \tag{8.24}$$

where ('+' corresponds to $\vartheta = -\beta$). The Melnikov vectors evaluated on these homoclinic orbits are not necessarily even and are in fact not even in n. For the purpose of calculating the Melnikov functions, only the even parts of the Melnikov vectors are needed, which are given by

$$\left.\frac{\delta \tilde{F}_1}{\delta \vec{Q}_n}\right|_{\text{even}} = \hat{K}^{(e)} \operatorname{sech}(2\mu t + 2p)(\Pi_n)^{-1} \begin{pmatrix} \hat{X}_n^{(1,e)} \\ -\hat{X}_n^{(2,e)} \end{pmatrix}, \tag{8.25}$$

where

$$\begin{aligned}
\hat{K}^{(e)} &= -2N(1-z^4)(8a\rho^{3/2}z^2)^{-1}\sqrt{\rho\cos^2\beta - 1}, \\
\Pi_n &= [\cos\beta \pm \cos\varphi \ \operatorname{sech}\ (2\mu t + 2p)\cos[2(n-1)\beta]] \\
&\quad \times [\cos\beta \pm \cos\varphi \ \operatorname{sech}\ (2\mu t + 2p)\cos[2(n+1)\beta]], \\
\hat{X}_n^{(1,e)} &= [\cos\beta \ \operatorname{sech}\ (2\mu t + 2p) \pm [\cos\varphi - i\sin\varphi \tanh\ (2\mu t + 2p)]\cos(2n\beta)]e^{i2\theta(t)}, \\
\hat{X}_n^{(2,e)} &= [\cos\beta \ \operatorname{sech}\ (2\mu t + 2p) \pm [\cos\varphi + i\sin\varphi \tanh\ (2\mu t + 2p)]\cos(2n\beta)]e^{-i2\theta(t)}.
\end{aligned}$$

The homoclinic orbit (8.22) represents the figure eight structure. If we denote by S the circle, we have the topological identification:

$$(\text{figure eight}) \ \otimes S = \bigcup_{p\in(-\infty,\infty),\ \gamma\in[0,2\pi]} Q_n(p, \gamma, a, \omega, \pm, N).$$

Chapter 9

Davey-Stewartson II Equation

9.1 Background

Davey-Stewartson Equation (DS) is a canonical high dimensional integrable system. It has a variety of applications. The original form was derived by A. Davey and K. Stewartson [30] as a model of water wave. It is this form that we shall focus on in this chapter. For more details on this chapter, see [62, 72].

Consider the Davey-Stewartson II equation (DSII)

$$\begin{cases} iq_t = \Box q + [2(|q|^2 - \omega^2) + u_y]q, \\ \Delta u = -4\partial_y |q|^2, \end{cases} \tag{9.1}$$

where q is a complex-valued function of the three variables (t, x, y), u is a real-valued function of the three variables (t, x, y), ω is a parameter, and

$$\Box = \partial_{xx} - \partial_{yy}, \quad \Delta = \partial_{xx} + \partial_{yy}, \quad i = \sqrt{-1}.$$

Periodic boundary condition is imposed,

$$q(t, x + 2\pi/\kappa_1, y) = q(t, x, y) = q(t, x, y + 2\pi/\kappa_2),$$
$$u(t, x + 2\pi/\kappa_1, y) = u(t, x, y) = u(t, x, y + 2\pi/\kappa_2),$$

where κ_1 and κ_2 are positive constants. Even constraint is also imposed,

$$q(t, -x, y) = q(t, x, y) = q(t, x, -y),$$
$$u(t, -x, y) = u(t, x, y) = u(t, x, -y).$$

The following constraint guarantees the existence of only two unstable modes,

$$\kappa_2 < \kappa_1 < 2\kappa_2, \text{ or } \kappa_1 < \kappa_2 < 2\kappa_1,$$

as shown in next section. The Davey-Stewartson equation is Gauge equivalent to the Ishimori equation [49] in the same way as the nonlinear Schrödinger equation is Gauge equivalent to the Heisenberg ferromagnet equation. Davey-Stewartson equation can be regarded as a higher dimensional generalization of the nonlinear Schrödinger equation studied in Chapter 3. In fact, it is a nontrivial generalization in the sense that the spatial part of the Lax pair of the DSII is a system of two first order partial differential equations, for which there is no convenient Floquet theory. It turns out that Melnikov vectors can still be obtained through quadratic products of Bloch eigenfunctions, instead of the gradient of the Floquet discriminant as in the NLS case.

At the moment, there is no global well-posedness for DSII in Sobolev spaces. In fact, DSII has finite-time blow-up solutions in $H^s(\mathbb{R}^2)$ $(0 < s < 1)$ [94].

9.2 Linear Stability

Consider the spatially uniform solution

$$q_c = \eta e^{-2i(\eta^2-\omega^2)t+i\gamma}, \tag{9.2}$$

where η is the amplitude and γ is the initial phase. Making the change of variable

$$q = e^{-2i(\eta^2-\omega^2)t+i\gamma}(\eta + \hat{q}),$$

and linearizing the DSII with regard to

$$\hat{q} = e^{i(k_1 x + k_2 y)+\Omega t}, \quad k_1, k_2 \in \mathbb{R},$$

we have

$$\Omega = \pm \frac{|k_1^2 - k_2^2|}{\sqrt{k_1^2 + k_2^2}} \sqrt{4\eta^2 - (k_1^2 + k_2^2)}.$$

Therefore, the unstable modes $(k_1, k_2) \in \mathbb{R}^2$ lie inside the circle $k_1^2 + k_2^2 = 4\eta^2$ and away from the lines $k_1 = k_2$ and $k_1 = -k_2$ on the plane.

9.3 Lax Pair and Darboux Transformations

The DSII has the Lax pair

$$L\psi = \lambda\psi, \tag{9.3}$$

$$\partial_t \psi = A\psi, \tag{9.4}$$

where $\psi = (\psi_1, \psi_2)$, and

$$L = \begin{pmatrix} D^- & q \\ \bar{q} & D^+ \end{pmatrix},$$

$$A = i\left[2\begin{pmatrix} -\partial_x^2 & q\partial_x \\ \bar{q}\partial_x & \partial_x^2 \end{pmatrix} + \begin{pmatrix} r_1 & (D^+q) \\ -(D^-\bar{q}) & r_2 \end{pmatrix}\right],$$

$$D^+ = \alpha\partial_y + \partial_x, \quad D^- = \alpha\partial_y - \partial_x, \quad \alpha^2 = -1. \tag{9.5}$$

r_1 and r_2 have the expressions,

$$r_1 = \frac{1}{2}(-w + iv), \quad r_2 = \frac{1}{2}(w + iv), \tag{9.6}$$

where u and v are real-valued functions satisfying

$$(\partial_x^2 + \partial_y^2)w = 2(\partial_x^2 - \partial_y^2)|q|^2, \tag{9.7}$$

$$(\partial_x^2 + \partial_y^2)v = i4\alpha\partial_x\partial_y|q|^2, \tag{9.8}$$

and $w = 2(|q|^2 - \omega^2) + u_y$. Notice that DSII (9.1) is invariant under the transformation σ:

$$\sigma \circ (q, \bar{q}, r_1, r_2; \alpha) = (q, \bar{q}, -r_2, -r_1; -\alpha). \tag{9.9}$$

Applying the transformation σ (9.9) to the Lax pair (9.3) and (9.4), we have a congruent Lax pair for which the compatibility condition gives the same DSII. The congruent Lax pair is given as:

$$\hat{L}\hat{\psi} = \lambda\hat{\psi}, \tag{9.10}$$

$$\partial_t\hat{\psi} = \hat{A}\hat{\psi}, \tag{9.11}$$

where $\hat{\psi} = (\hat{\psi}_1, \hat{\psi}_2)$, and

$$\hat{L} = \begin{pmatrix} -D^+ & q \\ \bar{q} & -D^- \end{pmatrix},$$

$$\hat{A} = i\left[2\begin{pmatrix} -\partial_x^2 & q\partial_x \\ \bar{q}\partial_x & \partial_x^2 \end{pmatrix} + \begin{pmatrix} -r_2 & -(D^-q) \\ (D^+\bar{q}) & -r_1 \end{pmatrix}\right].$$

The compatibility condition of the Lax pair (9.3) and (9.4),

$$\partial_t L = [A, L],$$

where $[A, L] = AL - LA$, and the compatibility condition of the congruent Lax pair (9.10, 9.11),

$$\partial_t\hat{L} = [\hat{A}, \hat{L}]$$

give the same DSII (9.1). Let (q, u) be a solution to the DSII (9.1), and let λ_0 be any value of λ. Let $\psi = (\psi_1, \psi_2)$ be a solution to the Lax pair (9.3) and (9.4) at $(q, \bar{q}, r_1, r_2; \lambda_0)$. Define the matrix operator:

$$\Gamma = \begin{bmatrix} \wedge + a & b \\ c & \wedge + d \end{bmatrix},$$

where $\wedge = \alpha\partial_y - \lambda$, and a, b, c, d are functions defined as:

$$a = \frac{1}{\Delta}(\psi_2 \wedge_2 \bar{\psi}_2 + \bar{\psi}_1 \wedge_1 \psi_1),$$
$$b = \frac{1}{\Delta}(\bar{\psi}_2 \wedge_1 \psi_1 - \psi_1 \wedge_2 \bar{\psi}_2),$$
$$c = \frac{1}{\Delta}(\bar{\psi}_1 \wedge_1 \psi_2 - \psi_2 \wedge_2 \bar{\psi}_1),$$
$$d = \frac{1}{\Delta}(\bar{\psi}_2 \wedge_1 \psi_2 + \psi_1 \wedge_2 \bar{\psi}_1),$$

in which $\wedge_1 = \alpha\partial_y - \lambda_0$, $\wedge_2 = \alpha\partial_y + \bar{\lambda}_0$, and

$$\Delta = -(|\psi_1|^2 + |\psi_2|^2).$$

Define a transformation as follows:

$$\begin{cases} (q, r_1, r_2) \to (Q, R_1, R_2), \\ \quad\quad \phi \quad \to \quad \Phi; \end{cases}$$

$$\begin{aligned} Q &= q - 2b, \\ R_1 &= r_1 + 2(D^+ a), \\ R_2 &= r_2 - 2(D^- d), \\ \Phi &= \Gamma\phi; \end{aligned} \tag{9.12}$$

where ϕ is any solution to the Lax pair (9.3) and (9.4) at $(q, \bar{q}, r_1, r_2; \lambda)$, D^+ and D^- are defined in (9.5), we have the following theorem [62].

Theorem 9.1. *The transformation* (9.12) *is a Bäcklund-Darboux transformation. That is, the function Q defined through the transformation* (9.12) *is also a solution to the DSII* (9.1)*. The function Φ defined through the transformation* (9.12) *solves the Lax pair* (9.3) *and* (9.4) *at* $(Q, \bar{Q}, R_1, R_2; \lambda)$.

9.4 Homoclinic Orbits

Since DSII is a higher dimensional generalization of NLS, a nontrivial homoclinic orbit should depend on both x and y. Consider the spatially independent solution,

$$q_c = \eta \exp[-2i(\eta^2 - \omega^2)t + i\gamma]. \tag{9.13}$$

As shown in Section 9.2, the dispersion relation for the linearized DSII at q_c is

$$\Omega = \pm \frac{|\xi_1^2 - \xi_2^2|}{\sqrt{\xi_1^2 + \xi_2^2}} \sqrt{4\eta^2 - (\xi_1^2 + \xi_2^2)}, \quad \text{for } \delta q \sim q_c \exp[i(\xi_1 x + \xi_2 y) + \Omega t],$$

where $\xi_1 = k_1\kappa_1$, $\xi_2 = k_2\kappa_2$, and k_1 and k_2 are integers. We restrict κ_1 and κ_2 as follows to have only two unstable modes $(\pm\kappa_1, 0)$ and $(0, \pm\kappa_2)$,

$$\kappa_2 < \kappa_1 < 2\kappa_2, \quad \kappa_1^2 < 4\eta^2 < \min\{\kappa_1^2 + \kappa_2^2, 4\kappa_2^2\},$$

or

$$\kappa_1 < \kappa_2 < 2\kappa_1, \quad \kappa_2^2 < 4\eta^2 < \min\{\kappa_1^2 + \kappa_2^2, 4\kappa_1^2\}.$$

The Bloch eigenfunction of the Lax pair (9.3) and (9.4) is given as,

$$\psi = c(t) \begin{bmatrix} -q_c \\ \chi \end{bmatrix} \exp[i(\xi_1 x + \xi_2 y)], \tag{9.14}$$

where

$$\begin{aligned} &c(t) = c_0 \exp\left\{[2\xi_1(i\alpha\xi_2 - \lambda) + ir_2]\, t\right\}, \\ &r_2 - r_1 = 2(|q_c|^2 - \omega^2), \\ &\chi = (i\alpha\xi_2 - \lambda) - i\xi_1, \\ &(i\alpha\xi_2 - \lambda)^2 + \xi_1^2 = \eta^2. \end{aligned}$$

For the iteration of the Bäcklund-Darboux transformations, one needs two sets of eigenfunctions. First, we choose $\xi_1 = \pm\frac{1}{2}\kappa_1$, $\xi_2 = 0$, $\lambda_0 = \sqrt{\eta^2 - \frac{1}{4}\kappa_1^2}$ (for a fixed branch),

$$\psi^\pm = c^\pm \begin{bmatrix} -q_c \\ \chi^\pm \end{bmatrix} \exp\left(\pm i\frac{1}{2}\kappa_1 x\right), \tag{9.15}$$

where

$$\begin{aligned} c^\pm &= c_0^\pm \exp\left\{(\mp\kappa_1\lambda_0 + ir_2)t\right\}, \\ \chi^\pm &= -\lambda_0 \mp i\frac{1}{2}\kappa_1 = \eta e^{\mp i(\frac{\pi}{2} + \vartheta_1)}. \end{aligned}$$

We apply the Bäcklund-Darboux transformations with $\psi = \psi^+ + \psi^-$, which generates the unstable foliation associated with the $(\kappa_1, 0)$ and $(-\kappa_1, 0)$ linearly unstable modes. Then, we choose $\xi_2 = \pm\frac{1}{2}\kappa_2$, $\lambda = 0$, $\xi_1^0 = \sqrt{\eta^2 - \frac{1}{4}\kappa_2^2}$ (for a fixed branch),

$$\phi_\pm = c_\pm \begin{bmatrix} -q_c \\ \chi_\pm \end{bmatrix} \exp\left[i\left(\xi_1^0 x \pm \frac{1}{2}\kappa_2 y\right)\right], \tag{9.16}$$

where

$$\begin{aligned} c_\pm &= c_\pm^0 \exp\left\{\left(\pm i\alpha\kappa_2\xi_1^0 + ir_2\right) t\right\}, \\ \chi_\pm &= \pm i\alpha\frac{1}{2}\kappa_2 - i\xi_1^0 = \pm\eta e^{\mp i\vartheta_2}. \end{aligned}$$

We start from these eigenfunctions $\phi_\pm$ to generate $\Gamma\phi_\pm$ through Bäcklund-Darboux transformations, and then iterate the Bäcklund-Darboux transformations with $\Gamma\phi_+ + \Gamma\phi_-$ to generate the unstable foliation associated with all the linearly unstable modes $(\pm\kappa_1, 0)$ and $(0, \pm\kappa_2)$. It turns out that the following representations are convenient,

$$\psi^\pm = \sqrt{c_0^+ c_0^-}\, e^{ir_2 t} \begin{pmatrix} v_1^\pm \\ v_2^\pm \end{pmatrix}, \tag{9.17}$$

$$\phi_\pm = \sqrt{c_+^0 c_-^0}\, e^{i\xi_1^0 x + ir_2 t} \begin{pmatrix} w_1^\pm \\ w_2^\pm \end{pmatrix}, \tag{9.18}$$

where

$$v_1^\pm = -q_c e^{\mp\frac{\tau}{2}\pm i\tilde{x}}, \quad v_2^\pm = \eta e^{\mp\frac{\tau}{2}\pm i\tilde{z}},$$
$$w_1^\pm = -q_c e^{\pm\frac{\hat{\tau}}{2}\pm i\hat{y}}, \quad w_2^\pm = \pm\eta e^{\pm\frac{\hat{\tau}}{2}\pm i\hat{z}},$$

and

$$c_0^+/c_0^- = e^{\rho+i\vartheta}, \quad \tau = 2\kappa_1\lambda_0 t - \rho, \quad \tilde{x} = \frac{1}{2}\kappa_1 x + \frac{\vartheta}{2}, \quad \tilde{z} = \tilde{x} - \frac{\pi}{2} - \vartheta_1,$$
$$c_+^0/c_-^0 = e^{\hat{\rho}+i\hat{\vartheta}}, \quad \hat{\tau} = 2i\alpha\kappa_2\xi_1^0 t + \hat{\rho}, \quad \hat{y} = \frac{1}{2}\kappa_2 y + \frac{\hat{\vartheta}}{2}, \quad \hat{z} = \hat{y} - \vartheta_2.$$

The following representations are also very useful,

$$\psi = \psi^+ + \psi^- = 2\sqrt{c_0^+ c_0^-}\, e^{ir_2 t} \begin{pmatrix} v_1 \\ v_2 \end{pmatrix}, \tag{9.19}$$

$$\phi = \phi^+ + \phi^- = 2\sqrt{c_+^0 c_-^0}\, e^{i\xi_1^0 x + ir_2 t} \begin{pmatrix} w_1 \\ w_2 \end{pmatrix}, \tag{9.20}$$

where

$$v_1 = -q_c\left(\cosh\frac{\tau}{2}\cos\tilde{x} - i\sinh\frac{\tau}{2}\sin\tilde{x}\right), \quad v_2 = \eta\left(\cosh\frac{\tau}{2}\cos\tilde{z} - i\sinh\frac{\tau}{2}\sin\tilde{z}\right),$$
$$w_1 = -q_c\left(\cosh\frac{\hat{\tau}}{2}\cos\hat{y} + i\sinh\frac{\hat{\tau}}{2}\sin\hat{y}\right), \quad w_2 = \eta\left(\sinh\frac{\hat{\tau}}{2}\cos\hat{z} + i\cosh\frac{\hat{\tau}}{2}\sin\hat{z}\right).$$

Applying the Bäcklund-Darboux transformations (9.12) with ψ given in (9.19), we have the representations,

$$a = -\lambda_0 \operatorname{sech}\tau \sin(\tilde{x}+\tilde{z})\sin(\tilde{x}-\tilde{z}) \tag{9.21}$$
$$\times[1 + \operatorname{sech}\tau\cos(\tilde{x}+\tilde{z})\cos(\tilde{x}-\tilde{z})]^{-1},$$

$$b = -q_c\tilde{b} = -\frac{\lambda_0 q_c}{\eta}[\cos(\tilde{x}-\tilde{z}) - i\tanh\tau\sin(\tilde{x}-\tilde{z}) \tag{9.22}$$
$$+ \operatorname{sech}\tau\cos(\tilde{x}+\tilde{z})][1 + \operatorname{sech}\tau\cos(\tilde{x}+\tilde{z})\cos(\tilde{x}-\tilde{z})]^{-1},$$

$$c = \bar{b}, \quad d = -\bar{a} = -a. \tag{9.23}$$

The evenness of b in x is enforced by the requirement that $\vartheta - \vartheta_1 = \pm\frac{\pi}{2}$, and

$$a^{\pm} = \mp\lambda_0 \text{ sech } \tau \cos\vartheta_1 \sin(\kappa_1 x) \times[1 \mp \text{ sech } \tau \sin\vartheta_1 \cos(\kappa_1 x)]^{-1}, \tag{9.24}$$

$$b^{\pm} = -q_c\tilde{b}^{\pm} = -\frac{\lambda_0 q_c}{\eta}[-\sin\vartheta_1 - i\tanh\tau\cos\vartheta_1 \pm \text{ sech } \tau\cos(\kappa_1 x)][1 \mp \text{ sech } \tau \sin\vartheta_1 \cos(\kappa_1 x)]^{-1}, \tag{9.25}$$

$$c = \overline{b}, \quad d = -\overline{a} = -a. \tag{9.26}$$

Notice also that $a^{\pm}$ is an odd function in x. Under the above Bäcklund-Darboux transformations, the eigenfunctions $\phi_\pm$ (9.16) are transformed into

$$\varphi^{\pm} = \Gamma\phi_\pm, \tag{9.27}$$

where

$$\Gamma = \begin{bmatrix} \Lambda + a & b \\ \overline{b} & \Lambda - a \end{bmatrix},$$

and $\Lambda = \alpha\partial_y - \lambda$ with λ evaluated at 0. Let $\varphi = \varphi^+ + \varphi^-$ (the arbitrary constants $c^0_\pm$ are already included in $\varphi^\pm$), φ has the representation,

$$\varphi = 2\sqrt{c^0_+ c^0_-}\, e^{i\xi^0_1 x + ir_2 t}\begin{bmatrix} -q_c W_1 \\ \eta W_2 \end{bmatrix}, \tag{9.28}$$

where

$$W_1 = (\alpha\partial_y w_1) + aw_1 + \eta\tilde{b}w_2, \quad W_2 = (\alpha\partial_y w_2) - aw_2 + \eta\overline{\tilde{b}}w_1.$$

We generate the coefficients in the Bäcklund-Darboux transformations (9.12) with φ (the iteration of the Bäcklund-Darboux transformations),

$$a^{(I)} = -[W_2(\alpha\partial_y\overline{W_2}) + \overline{W_1}(\alpha\partial_y W_1)](|W_1|^2 + |W_2|^2)^{-1}, \tag{9.29}$$

$$b^{(I)} = \frac{q_c}{\eta}[\overline{W_2}(\alpha\partial_y W_1) - W_1(\alpha\partial_y\overline{W_2})](|W_1|^2 + |W_2|^2)^{-1}, \tag{9.30}$$

$$c^{(I)} = \overline{b^{(I)}}, \quad d^{(I)} = -\overline{a^{(I)}}, \tag{9.31}$$

where

$$W_2(\alpha\partial_y\overline{W_2}) + \overline{W_1}(\alpha\partial_y W_1) = \frac{1}{2}\alpha\kappa_2\Big\{\cosh\hat{\tau}\Big[-\alpha\kappa_2 a + ia\eta(\tilde{b} + \overline{\tilde{b}})\cos\vartheta_2\Big] + \Big[\frac{1}{4}\kappa_2^2 - a^2 - \eta^2|\tilde{b}|^2\Big]\cos(\hat{y} + \hat{z})\sin\vartheta_2 + \sinh\hat{\tau}\Big[a\eta(\tilde{b} - \overline{\tilde{b}})\sin\vartheta_2\Big]\Big\},$$

$$
\begin{aligned}
&|W_1|^2 + |W_2|^2 \\
&= \cosh\hat{\tau}\left[a^2 + \frac{1}{4}\kappa_2^2 + \eta^2|\tilde{b}|^2 + i\alpha\kappa_2\eta\frac{1}{2}(\tilde{b}+\bar{\tilde{b}})\cos\vartheta_2\right] \\
&\quad + \left[\frac{1}{4}\kappa_2^2 - a^2 - \eta^2|\tilde{b}|^2\right]\sin(\hat{y}+\hat{z})\sin\vartheta_2 + \sinh\hat{\tau}\left[\alpha\kappa_2\eta\frac{1}{2}(\tilde{b}-\bar{\tilde{b}})\sin\vartheta_2\right], \\
&\overline{W_2}(\alpha\partial_y W_1) - W_1(\alpha\partial_y \overline{W_2}) \\
&= \frac{1}{2}\alpha\kappa_2\left\{\cosh\hat{\tau}\left[-\alpha\kappa_2\eta\tilde{b} + i(-a^2 + \frac{1}{4}\kappa_2^2 + \eta^2\tilde{b}^2)\cos\vartheta_2\right]\right. \\
&\quad \left. + \sinh\hat{\tau}(a^2 - \frac{1}{4}\kappa_2^2 + \eta^2\tilde{b}^2)\sin\vartheta_2\right\}.
\end{aligned}
$$

The new solution to the focusing Davey-Stewartson II equation (9.1) is given by

$$Q = q_c - 2b - 2b^{(I)}. \tag{9.32}$$

The evenness of $b^{(I)}$ in y is enforced by the requirement that $\hat{\vartheta} - \vartheta_2 = \pm\frac{\pi}{2}$. In fact, we have

Lemma 9.2. *Under the requirements that $\vartheta - \vartheta_1 = \pm\frac{\pi}{2}$, and $\hat{\vartheta} - \vartheta_2 = \pm\frac{\pi}{2}$,*

$$b(-x) = b(x), \quad b^{(I)}(-x,y) = b^{(I)}(x,y) = b^{(I)}(x,-y), \tag{9.33}$$

and $Q = q_c - 2b - 2b^{(I)}$ is even in both x and y.

Proof: It is a direct verification by noticing that under the requirements, a is an odd function in x. □

The asymptotic behavior of Q can be computed directly. In fact, we have the asymptotic phase shift lemma.

Lemma 9.3 (Asymptotic Phase Shift Lemma). *For $\lambda_0 > 0$, $\xi_1^0 > 0$, and $i\alpha = 1$, as $t \to \pm\infty$,*

$$Q = q_c - 2b - 2b^{(I)} \to q_c e^{i\pi} e^{\mp i2(\vartheta_1 - \vartheta_2)}. \tag{9.34}$$

In comparison, the asymptotic phase shift of the first application of the Bäcklund-Darboux transformations is given by

$$q_c - 2b \to q_c e^{\mp i2\vartheta_1}.$$

9.5 Melnikov Vectors

The DSII (9.1) can be written in the Hamiltonian form,

$$\begin{cases} iq_t = \delta H/\delta\bar{q}, \\ i\bar{q}_t = -\delta H/\delta q, \end{cases} \tag{9.35}$$

where

$$H = \int_0^{2\pi} \int_0^{2\pi} \left[|q_y|^2 - |q_x|^2 + \frac{1}{2}(r_2 - r_1) |q|^2 \right] dx\, dy.$$

We have the lemma [62].

Lemma 9.4. *The inner product of the vector*

$$\mathcal{U} = \begin{pmatrix} \psi_2 \hat{\psi}_2 \\ \psi_1 \hat{\psi}_1 \end{pmatrix}^- + S \begin{pmatrix} \psi_2 \hat{\psi}_2 \\ \psi_1 \hat{\psi}_1 \end{pmatrix},$$

where $\psi = (\psi_1, \psi_2)$ *is an eigenfunction solving the Lax pair* (9.3) *and* (9.4), *and* $\hat{\psi} = (\hat{\psi}_1, \hat{\psi}_2)$ *is an eigenfunction solving the corresponding congruent Lax pair* (9.10) *and* (9.11), *and* $S = \begin{pmatrix} 0 & 1 \\ 1 & 0 \end{pmatrix}$, *with the vector field* $J\nabla H$ *given by the right hand side of* (9.35) *vanishes,*

$$\langle \mathcal{U},\, J\nabla H \rangle = 0,$$

where $J = \begin{pmatrix} 0 & 1 \\ -1 & 0 \end{pmatrix}$.

If we only consider even functions, i.e., q and $u = r_2 - r_1$ are even functions in both x and y, then we can split $\mathcal{U}$ into its even and odd parts,

$$\mathcal{U} = \mathcal{U}^{(e,x)}_{(e,y)} + \mathcal{U}^{(e,x)}_{(o,y)} + \mathcal{U}^{(o,x)}_{(e,y)} + \mathcal{U}^{(o,x)}_{(o,y)},$$

where

$$\begin{aligned}
\mathcal{U}^{(e,x)}_{(e,y)} &= \frac{1}{4}[\mathcal{U}(x,y) + \mathcal{U}(-x,y) + \mathcal{U}(x,-y) + \mathcal{U}(-x,-y)], \\
\mathcal{U}^{(e,x)}_{(o,y)} &= \frac{1}{4}[\mathcal{U}(x,y) + \mathcal{U}(-x,y) - \mathcal{U}(x,-y) - \mathcal{U}(-x,-y)], \\
\mathcal{U}^{(o,x)}_{(e,y)} &= \frac{1}{4}[\mathcal{U}(x,y) - \mathcal{U}(-x,y) + \mathcal{U}(x,-y) - \mathcal{U}(-x,-y)], \\
\mathcal{U}^{(o,x)}_{(o,y)} &= \frac{1}{4}[\mathcal{U}(x,y) - \mathcal{U}(-x,y) - \mathcal{U}(x,-y) + \mathcal{U}(-x,-y)].
\end{aligned}$$

Then we have the lemma [62].

Lemma 9.5. *When* q *and* $u = r_2 - r_1$ *are even functions in both* x *and* y, *we have*

$$\langle \mathcal{U}^{(e,x)}_{(e,y)},\, J\nabla H \rangle = 0.$$

9.5.1 Melnikov Integrals

Consider the perturbed DSII equation,

$$\begin{cases} i\partial_t q = [\partial_x^2 - \partial_y^2] q + [2(|q|^2 - \omega^2) + u_y] q + \epsilon i f, \\ [\partial_x^2 + \partial_y^2] u = -4\partial_y |q|^2, \end{cases} \tag{9.36}$$

where q and u are respectively complex-valued and real-valued functions of three variables (t, x, y), and $G = (f, \overline{f})$ are the perturbation terms which can depend on q and $\overline{q}$ and their derivatives and t, x and y. The Melnikov integral is given by [62],

$$\begin{aligned} M &= \int_{-\infty}^{\infty} \langle \mathcal{U}, G \rangle \, dt \\ &= 2 \int_{-\infty}^{\infty} \int_0^{2\pi} \int_0^{2\pi} R_e[(\psi_2 \hat{\psi}_2) f + (\psi_1 \hat{\psi}_1) \overline{f}] \, dx \, dy \, dt, \end{aligned} \tag{9.37}$$

where the integrand is evaluated on an unperturbed homoclinic orbit in certain center-unstable (= center-stable) manifold, and such orbit can be obtained through the Bäcklund-Darboux transformations given in Theorem 9.1. When we only consider even functions, i.e., q and u are even functions in both x and y, the corresponding Melnikov function is given by [62],

$$\begin{aligned} M^{(e)} &= \int_{-\infty}^{\infty} \langle \mathcal{U}^{(e,x)}_{(e,y)}, \vec{G} \rangle \, dt \\ &= \int_{-\infty}^{\infty} \langle \mathcal{U}, \vec{G} \rangle \, dt, \end{aligned} \tag{9.38}$$

which is the same as expression (9.37).

9.5.2 An Example

We continue from the example in Section 9.4. We generate the following eigenfunctions corresponding to the potential Q given in (9.32) through the iterated Bäcklund-Darboux transformations,

$$\Psi^{\pm} = \Gamma^{(I)} \Gamma \psi^{\pm}, \quad \text{at} \quad \lambda = \lambda_0 = \sqrt{\eta^2 - \frac{1}{4}\kappa_1^2}, \tag{9.39}$$

$$\Phi_{\pm} = \Gamma^{(I)} \Gamma \phi_{\pm}, \quad \text{at} \quad \lambda = 0, \tag{9.40}$$

where

$$\Gamma = \begin{bmatrix} \Lambda + a & b \\ \overline{b} & \Lambda - a \end{bmatrix}, \quad \Gamma^{(I)} = \begin{bmatrix} \Lambda + a^{(I)} & b^{(I)} \\ \overline{b^{(I)}} & \Lambda - \overline{a^{(I)}} \end{bmatrix},$$

where $\Lambda = \alpha \partial_y - \lambda$ for general λ.

Lemma 9.6 (see [62, 72]). *The eigenfunctions $\Psi^{\pm}$ and $\Phi_{\pm}$ defined in (9.39) and (9.40) have the representations,*

$$\Psi^{\pm} = \frac{\pm 2\lambda_0 W(\psi^+, \psi^-)}{\Delta} \begin{bmatrix} (-\lambda_0 + a^{(I)})\overline{\psi}_2 - b^{(I)}\overline{\psi}_1 \\ \overline{b^{(I)}}\overline{\psi}_2 + (\lambda_0 + \overline{a^{(I)}})\overline{\psi}_1 \end{bmatrix}, \tag{9.41}$$

$$\Phi_{\pm} = \frac{\mp i \alpha \kappa_2}{\Delta^{(I)}} \begin{bmatrix} \Xi_1 \\ \Xi_2 \end{bmatrix}, \tag{9.42}$$

where

$$W(\psi^+,\psi^-) = \begin{vmatrix} \psi_1^+ & \psi_1^- \\ \psi_2^+ & \psi_2^- \end{vmatrix} = -i\kappa_1 c_0^+ c_0^- q_c \exp(i2r_2 t),$$
$$\Delta = -(|\psi_1|^2 + |\psi_2|^2),$$
$$\psi = \psi^+ + \psi^-,$$
$$\Delta^{(I)} = -(|\varphi_1|^2 + |\varphi_2|^2),$$
$$\varphi = \varphi^+ + \varphi^-,$$
$$\varphi^\pm = \Gamma\phi_\pm \text{ at } \lambda = 0,$$
$$\Xi_1 = \overline{\varphi}_1(\varphi_1^+\varphi_1^-) + \overline{\varphi_2^+}(\varphi_1^+\varphi_2^-) + \overline{\varphi_2^-}(\varphi_1^-\varphi_2^+),$$
$$\Xi_2 = \overline{\varphi}_2(\varphi_2^+\varphi_2^-) + \overline{\varphi_1^+}(\varphi_1^-\varphi_2^+) + \overline{\varphi_1^-}(\varphi_1^+\varphi_2^-).$$

If we take r_2 *to be real (in the Melnikov vectors,* r_2 *appears in the form* $r_2 - r_1 = 2(|q_c|^2 - \omega^2)$*), then*

$$\Psi^\pm \longrightarrow 0, \quad \Phi_\pm \longrightarrow 0, \quad \text{as } t \to \pm\infty. \tag{9.43}$$

Next we generate eigenfunctions solving the corresponding congruent Lax pair (9.10) and (9.11) with the potential Q, through the iterated Bäcklund-Darboux transformations and the symmetry transformation (9.9) [62, 72].

Lemma 9.7. *Under the replacements*

$$\alpha \longrightarrow -\alpha \quad (\vartheta_2 \longrightarrow \pi - \vartheta_2), \quad \hat{\vartheta} \longrightarrow \hat{\vartheta} + \pi - 2\vartheta_2, \quad \hat{\rho} \longrightarrow -\hat{\rho},$$

the coefficients in the iterated Bäcklund-Darboux transformations are transformed as follows,

$$a^{(I)} \longrightarrow \overline{a^{(I)}}, \quad b^{(I)} \longrightarrow b^{(I)},$$
$$(c^{(I)} = \overline{b^{(I)}}) \longrightarrow (c^{(I)} = \overline{b^{(I)}}), \quad (d^{(I)} = -\overline{a^{(I)}}) \longrightarrow (\overline{d^{(I)}} = -a^{(I)}).$$

Lemma 9.8 (see [62, 72]). *Under the replacements*

$$\begin{aligned} &\alpha \mapsto -\alpha \quad (\vartheta_2 \longrightarrow \pi - \vartheta_2), \quad r_1 \mapsto -r_2, \\ &r_2 \mapsto -r_1, \quad \hat{\vartheta} \longrightarrow \hat{\vartheta} + \pi - 2\vartheta_2, \quad \hat{\rho} \longrightarrow -\hat{\rho}, \end{aligned} \tag{9.44}$$

the potentials are transformed as follows,

$$Q \longrightarrow Q,$$
$$(R = \overline{Q}) \longrightarrow (R = \overline{Q}),$$
$$R_1 \longrightarrow -R_2,$$
$$R_2 \longrightarrow -R_1.$$

The eigenfunctions $\Psi^\pm$ and $\Phi_\pm$ given in (9.41) and (9.42) depend on the variables in the replacements (9.44):

$$\Psi^\pm = \Psi^\pm(\alpha, r_1, r_2, \hat{\vartheta}, \hat{\rho}),$$
$$\Phi_\pm = \Phi_\pm(\alpha, r_1, r_2, \hat{\vartheta}, \hat{\rho}).$$

Under replacements (9.44), $\Psi^\pm$ and $\Phi_\pm$ are transformed into

$$\widehat{\Psi}^\pm = \Psi^\pm(-\alpha, -r_2, -r_1, \hat{\vartheta} + \pi - 2\vartheta_2, -\hat{\rho}), \tag{9.45}$$
$$\widehat{\Phi}_\pm = \Phi_\pm(-\alpha, -r_2, -r_1, \hat{\vartheta} + \pi - 2\vartheta_2, -\hat{\rho}). \tag{9.46}$$

Corollary 9.9 (see [62, 72]). *$\widehat{\Psi}^\pm$ and $\widehat{\Phi}_\pm$ solve the congruent Lax pair* (9.10) *and* (9.11) *at* $(Q, \overline{Q}, R_1, R_2; \lambda_0)$ *and* $(Q, \overline{Q}, R_1, R_2; 0)$, *respectively.*

Notice that as a function of η, ξ_1^0 has two (plus and minus) branches. In order to construct Melnikov vectors, we need to study the effect of the replacement $\xi_1^0 \longrightarrow -\xi_1^0$.

Lemma 9.10 (see [62, 72]). *Under the replacements*

$$\xi_1^0 \longrightarrow -\xi_1^0 \quad (\vartheta_2 \longrightarrow -\vartheta_2), \quad \hat{\vartheta} \longrightarrow \hat{\vartheta} + \pi - 2\vartheta_2, \quad \hat{\rho} \longrightarrow -\hat{\rho}, \tag{9.47}$$

the coefficients in the iterated Bäcklund-Darboux transformations are invariant,

$$a^{(I)} \mapsto a^{(I)}, \quad b^{(I)} \mapsto b^{(I)},$$
$$(c^{(I)} = \overline{b^{(I)}}) \mapsto (c^{(I)} = \overline{b^{(I)}}), \quad (d^{(I)} = -\overline{a^{(I)}}) \mapsto (d^{(I)} = -\overline{a^{(I)}});$$

thus the potentials are also invariant,

$$Q \longrightarrow Q, \quad (R = \overline{Q}) \longrightarrow (R = \overline{Q}),$$
$$R_1 \longrightarrow R_1, \quad R_2 \longrightarrow R_2.$$

The eigenfunction $\Phi_\pm$ given in (9.42) depends on the variables in the replacements (9.47):

$$\Phi_\pm = \Phi_\pm(\xi_1^0, \hat{\vartheta}, \hat{\rho}).$$

Under the replacements (9.47), $\Phi_\pm$ is transformed into

$$\widetilde{\Phi}_\pm = \Phi_\pm(-\xi_1^0, \hat{\vartheta} + \pi - 2\vartheta_2, -\hat{\rho}). \tag{9.48}$$

Corollary 9.11. *$\widetilde{\Phi}_\pm$ solves the Lax pair* (9.3) *and* (9.4) *at* $(Q, \overline{Q}, R_1, R_2; 0)$.

In the construction of the Melnikov vectors, we need to replace $\Phi_\pm$ by $\widetilde{\Phi}_\pm$ to guarantee the periodicity in x of period $L_1 = \frac{2\pi}{\kappa_1}$.

The Melnikov vectors for the Davey-Stewartson II equations are given by,

$$\mathcal{U}^\pm = \begin{pmatrix} \Psi_2^\pm \widehat{\Psi}_2^\pm \\ \Psi_1^\pm \widehat{\Psi}_1^\pm \end{pmatrix}^- + S \begin{pmatrix} \Psi_2^\pm \widehat{\Psi}_2^\pm \\ \Psi_1^\pm \widehat{\Psi}_1^\pm \end{pmatrix}, \tag{9.49}$$

$$\mathcal{U}_\pm = \begin{pmatrix} \widetilde{\Phi}_\pm^{(2)} \widehat{\Phi}_\pm^{(2)} \\ \widetilde{\Phi}_\pm^{(1)} \widehat{\Phi}_\pm^{(1)} \end{pmatrix}^- + S \begin{pmatrix} \widetilde{\Phi}_\pm^{(2)} \widehat{\Phi}_\pm^{(2)} \\ \widetilde{\Phi}_\pm^{(1)} \widehat{\Phi}_\pm^{(1)} \end{pmatrix}, \tag{9.50}$$

where $S = \begin{pmatrix} 0 & 1 \\ 1 & 0 \end{pmatrix}$. The corresponding Melnikov functions (9.37) are given by,

$$\begin{aligned} M^{\pm} &= \int_{-\infty}^{\infty} \langle \mathcal{U}^{\pm}, \vec{G} \rangle \, dt \\ &= 2 \int_{-\infty}^{\infty} \int_{0}^{2\pi} \int_{0}^{2\pi} R_e[(\Psi_2^{\pm} \widehat{\Psi}_2^{\pm}) f(Q, \overline{Q}) \\ &\quad + (\Psi_1^{\pm} \widehat{\Psi}_1^{\pm}) \overline{f}(Q, \overline{Q})] \, dx \, dy \, dt, \end{aligned} \tag{9.51}$$

$$\begin{aligned} M_{\pm} &= \int_{-\infty}^{\infty} \langle \mathcal{U}_{\pm}, \vec{G} \rangle \, dt \\ &= 2 \int_{-\infty}^{\infty} \int_{0}^{2\pi} \int_{0}^{2\pi} R_e[(\widetilde{\Phi}_{\pm}^{(2)} \widehat{\Phi}_{\pm}^{(2)}) f(Q, \overline{Q}) \\ &\quad + (\widetilde{\Phi}_{\pm}^{(1)} \widehat{\Phi}_{\pm}^{(1)}) \overline{f}(Q, \overline{Q})] \, dx \, dy \, dt, \end{aligned} \tag{9.52}$$

where Q is given in (9.32), $\Psi^{\pm}$ is given in (9.41), $\widetilde{\Phi}_{\pm}$ is given in (9.42) and (9.48), $\widehat{\Psi}^{\pm}$ is given in (9.41) and (9.45), and $\widehat{\Phi}_{\pm}$ is given in (9.42) and (9.46). As given in (9.38), the above formulas also apply when we consider even function Q in both x and y.

9.6 Extra Comments

In general, one can classify soliton equations into two categories. Category I consists of those equations possessing instabilities, under periodic boundary condition. In their phase space, figure-eight structures (i.e., separatrices) exist. Category II consists of those equations possessing no instability, under periodic boundary condition. In their phase space, no figure-eight structure (i.e., separatrix) exists. Typical Category I soliton equations are for example, focusing nonlinear Schrödinger equation, sine-Gordon equation, modified KdV equation, Heisenberg ferromagnet equation, derivative nonlinear Schrödinger equation, discrete nonlinear Schrödinger equation, Davey-Stewartson II equation etc. Typical Category II soliton equations are for example, KdV equation, defocusing nonlinear Schrödinger equation, sinh-Gordon equation, Toda lattice, Boussinesq equation, KP equations etc. In principle, figure-eight structures for Category I soliton equations can be constructed through Darboux transformations, as illustrated in previous chapters. It should be remarked that Darboux transformations still exist for Category II soliton equations, but do not produce any figure-eight structure.

Chapter 10

Acoustic Spectral Problem

This chapter deals with the Darboux transformations for the acoustic problem and the related Harry Dym (HD) equation. The acoustic problem and the linear Schrödinger operator are closely connected. This connection is the key for studying the acoustic problem and the HD equation, see e.g., [31, 32, 44, 110]. However, as discussed below, the connection is not simple. This makes the acoustic problem interesting in its own right.

10.1 Physical Background

The acoustic problem describes wave propagation in an inhomogeneous media. The acoustic problem on the real line is described by the following ODE:

$$\psi_{xx} = \frac{\lambda}{u^2(x)}\psi, \tag{10.1}$$

where ψ is complex-valued, $u(x)$ is real-valued, and λ is a complex parameter in general. Here is one way how this equation arises in physics. Consider the Maxwell equations in a medium without external sources, with the standard notations $(\mathbf{E}, \mathbf{H})$ for the electromagnetic field and $\mathbf{D} = \epsilon\mathbf{E}$, $\mathbf{B} = \kappa\mathbf{H}$. Assume that the medium is isotropic but inhomogeneous, i.e., ϵ, κ are scalar quantities; and $\kappa \equiv 1$ and $\epsilon = \epsilon(x, y, z)$, then

$$[\nabla, \mathbf{B}] = \frac{1}{c}\frac{\partial \mathbf{D}}{\partial t}, \quad [\nabla, \mathbf{E}] = -\frac{1}{c}\frac{\partial \mathbf{B}}{\partial t}, \quad (\nabla, \mathbf{D}) = (\nabla, \mathbf{B}) = 0, \tag{10.2}$$

where $[\cdot,\cdot]$ and $(\cdot,\cdot)$ denote the vector and dot product in $\mathbb{R}^3$, respectively. Eliminating $\mathbf{B}$ from (10.2) leads to an equation connecting the quantities $\mathbf{E}$ and $\mathbf{D}$:

$$[\nabla, [\nabla, \mathbf{E}]] = -\frac{1}{c^2}\frac{\partial^2 \mathbf{D}}{\partial t^2}. \tag{10.3}$$

For the electric field $\mathbf{E}$, one seeks $\mathbf{E}(t;x,y,z) = e^{i\omega t}\psi(x,y,z)$ and takes into account the last equation of (10.2) to obtain

$$\nabla\left(\frac{(\psi,\nabla)\epsilon}{\epsilon}\right) + \Delta\psi = -\frac{\omega^2}{c^2}\epsilon\psi, \tag{10.4}$$

where Δ is a three dimensional Laplacian. Equation (10.1) follows if one lets $\epsilon = \epsilon(x)$, $\psi = (0,0,\psi(x))$, $\lambda = -\omega^2/c^2$, $u^{-2}(x) = \epsilon(x)$. The dielectric permittivity $\epsilon(x)$ will be henceforth referred to as a potential. Alternatively, one can choose $\epsilon = \epsilon(x,y)$ as well as $\psi = (0,0,\psi(x,y))$. If this is the case, (10.4) is reduced to a linear PDE

$$\Delta\psi = \lambda\epsilon\psi, \tag{10.5}$$

where Δ is a two dimensional Laplacian.

10.2 Connection with Linear Schrödinger Operator

Starting from (10.1), let [31, 32]

$$u(x) = v_y(y), \quad x = v(y). \tag{10.6}$$

This reduces equation (10.1) to

$$\psi_{yy} = U\psi_y + \lambda\psi, \tag{10.7}$$

with $U = v_{yy}/v_y$. Finally, a formal substitution

$$\psi \to \sqrt{v_y}\psi,$$

transforms (10.7) into the stationary Schrödinger equation

$$\psi_{yy} = (\lambda + V(y))\,\psi, \tag{10.8}$$

where the potential $V(y)$ is related to the potential $U(y)$ via

$$V = \frac{U^2 - 2U_y}{4}. \tag{10.9}$$

Notice also that with a substitution $U = -2p_y/p$, $p(y)$ in turn satisfies (10.8) with $\lambda = 0$.

10.3 Discrete Symmetries of the Acoustic Problem

Following Shabat [97], we seek elementary discrete symmetries of equation (10.7) in the form

$$\psi \to \psi^{(1)} = f\psi_y + g\psi, \tag{10.10}$$

for some λ-independent functions f and g of y. One easily verifies that there are three discrete symmetries of the type (10.10) for equation (10.7). They are

$$\psi \to \psi^{(1)} = \frac{\psi}{v}, \quad v \to v^{(1)} = \frac{1}{v}; \quad \psi \to \psi^{(1)} = \frac{\psi_y}{v_y}, \quad v \to v^{(1)} = \int \frac{dy}{v_y}; \tag{10.11}$$

and

$$\begin{gathered} \psi \to \psi^{(1)} = \frac{\psi_1 \psi_y}{\psi_{1,y}} - \psi, \\ v_y \to v_y^{(1)} = v_y \left(\frac{\psi_1}{\psi_{1,y}} \right)^2, \quad U \to U^{(1)} = U + 2D \ln \frac{\psi_1}{\psi_{1,y}}. \end{gathered} \tag{10.12}$$

In the latter equation $\psi_1 = \psi_1(y, \lambda_1)$ is a particular solution of (10.7) with the spectral parameter value λ_1, further referred to as a *prop solution*, $D = \partial_y$ and $\psi_{1,y} = D\psi_1$. The former two symmetries (10.11) define the new quantity $U^{(1)} = v_{yy}^{(1)}/v_y^{(1)}$ in a way independent of any solution $\psi(y, \lambda)$ of (10.7). These symmetries arise as a particular case of (10.10) as the result of gauging corresponding to the choice of f or g alternatively zero. These symmetries have a trivial kernel in the solution space of (10.7). According to the terminology of [97], we call the symmetries (10.11) T-symmetries, sometimes referred to as Schlesinger transforms[1]. On the other hand, the transformation (10.12) alias the Darboux transformation, which [97] calls an S-symmetry, does have a non-trivial kernel on the solution space of (10.7). This property is essential. We use the common term *dressing* for the application of the transformation (10.12) to a triplet (ψ, v, U), and the resulting triplet $(\psi^{(1)}, v^{(1)}, U^{(1)})$ is referred to as the dressed one.

10.4 Crum Formulae and Dressing Chains for the Acoustic Problem

In this section, we present the formulae describing an n-step dressing procedure for any $n \in \mathbb{N}$, analogues of which are known for the Schrödinger equation as the Crum formulae [28, 108]. A single act of dressing (10.12) can be iterated n times to yield a triple $(\psi^{(n)}, v^{(n)}, U^{(n)})$. One starts by dressing a triple

[1] In the context of soliton solutions, the T-symmetries play the part of explicitly invertible Bäcklund transforms, see e.g., [101].

$(\psi, v, U) = (\psi^{(0)}, v^{(0)}, U^{(0)})$ with a prop function $\psi_1^{(0)}$, which solves (10.7) at $(\lambda_1, U^{(0)})$. The resulting solution $\psi^{(1)}$ solves (10.7) with the dressed potential $U^{(1)}$. In the jth step ($j = 1, \ldots, n$), one uses a prop solution $\psi_j^{(j-1)}$, which solves (10.7) at $(\lambda_j, U^{(j-1)})$, to produce the j-times dressed solution $\psi^{(j)}$ and potential $U^{(j)}$ (as well as the function $v^{(j)}$ with $U^{(j)} = v_{yy}^{(j)}/v_y^{(j)}$). Note that the spectral parameter λ in the dressed equations for $\psi^{(j)}$ is the same for all $j = 1, \ldots, n$. It is easy to see that the n-times dressed solution $\psi^{(n)}$ shall have the form

$$\psi^{(n)} = \sum_{j=1}^{n} a_j D^j \psi + (-1)^n \psi, \tag{10.13}$$

with the function-coefficients a_j to be found, which of course will depend on the choice of the prop solutions $\psi_j^{(j-1)}$. It follows from (10.12) that

$$U^{(n)} = U + 2D \ln a_n, \tag{10.14}$$

thus

$$\psi^{(n)} = \prod_{j=1}^{n} \frac{\psi_j^{(j-1)}}{D\psi_j^{(j-1)}} D^n \psi^{(0)} + \cdots + (-1)^n \psi^{(0)}, \quad U^{(n)} = U^{(0)} + 2D \ln \prod_{j=1}^{n} \frac{\psi_j^{(j-1)}}{D\psi_j^{(j-1)}}.$$

So far the choice of the prop solutions $\psi_i^{(i-1)}$ has been quite arbitrary. But suppose now that the original equation (10.7) possesses n distinct formal solutions ψ_j, corresponding to spectral parameter values λ_j, $j = 1, \ldots, n$. Let $\psi_j \equiv \psi_j^{(0)}$ and consider the following dressing procedure:

$$\begin{array}{ccccccc}
\psi^{(0)} & \psi_1^{(0)} & \psi_2^{(0)} & \cdots & \psi_{n-1}^{(0)} & \psi_n^{(0)} & U^{(0)} \\
\psi^{(1)} & 0 & \psi_2^{(1)} & \cdots & \psi_{n-1}^{(1)} & \psi_n^{(1)} & U^{(1)} \\
\vdots & & & & & & \\
\psi^{(n-1)} & 0 & 0 & \cdots & 0 & \psi_n^{(n-1)} & U^{(n-1)} \\
\psi^{(n)} & 0 & 0 & \cdots & 0 & 0 & U^{(n)}
\end{array} \tag{10.15}$$

That is, for $j = 1, \ldots, n$ on the above diagram (10.15), each new line $j + 1$ is obtained by dressing the functions from the preceding line j by (10.12) with a prop solution $\psi_{\mathbf{j}}^{(\mathbf{j}-1)}$.

Zeroes, proliferating as one moves down the diagram stem from the non-trivial kernel property of the S-symmetry, and it is this property that now enables one to find the unknown functions a_j. Indeed, substitution of any $\psi_j = \psi_j^{(0)}$ for ψ in the right hand side of (10.13) shall yield zero. Hence, the coefficients a_j satisfy a system of n independent linear algebraic equations, namely

$$\sum_{k=1}^{n} a_k D^k \psi_j + (-1)^n \psi_j = 0, \quad j = 1, \ldots, n.$$

Solving it by the Cramer rule and substituting the result into (10.13) and (10.14), we obtain

$$U^{(n)} = U + 2D \ln \frac{\widetilde{\Delta}_n}{\Delta_n},$$

$$v_y^{(n)} = v_y \left(\frac{\widetilde{\Delta}_n}{\Delta_n}\right)^2, \quad \psi^{(n)} = \frac{\widehat{\Delta}_{n+1}}{\Delta_n}, \tag{10.16}$$

where Δ_n, $\widetilde{\Delta}_n$ are determinants of square $n \times n$ matrices, whereas $\widehat{\Delta}_{n+1}$ – of an $(n+1) \times (n+1)$ matrix as follows:

$$\Delta_n = \begin{vmatrix} D\psi_1 & \dots & D^n\psi_1 \\ \vdots & & \vdots \\ D\psi_n & \dots & D^n\psi_n \end{vmatrix}, \quad \widetilde{\Delta}_n = \begin{vmatrix} \psi_1 & \dots & D^{n-1}\psi_1 \\ \vdots & & \vdots \\ \psi_n & \dots & D^{n-1}\psi_n \end{vmatrix},$$

$$\widehat{\Delta}_{n+1} = \begin{vmatrix} \psi & \dots & D^n\psi \\ \psi_1 & \dots & D^n\psi_1 \\ \vdots & & \vdots \\ \psi_n & \dots & D^n\psi_n \end{vmatrix}. \tag{10.17}$$

Notice that the linearity of (10.7) makes the choice of the sign before $\psi^{(n)}$ irrelevant. The formulae (10.16) and (10.17) make it possible to find rich families of exact solutions of (10.1).

Example 10.1 (reflexionless potentials)**.** The easiest case is dressing from the birthday suit, or on the vacuum background, assuming $U = 0$ (hence $x = c_1 y + c_2$, $u(x) = c_1$, where c_1 and c_2 are arbitrary constants). Such a natural rigging yields gratis all the B-potentials reported by Novikov [93], moreover the formulae for their computation derived therein turn out to be particular cases of (10.12) and (10.13) with merely $v_y = 1$. For instance $\psi_1 = \sinh \xi_1$, $\psi_2 = \cosh \xi_2$ (here $n = 2$, y_j are constants, $\xi_i = k_j(y - y_j)$, $j = 1, 2$) yield a reflexionless B-potential with a power $2/3$ singularity: $\epsilon^{(2)} = \frac{1}{[u^{(2)}]^2}$, where[2]

$$u^{(2)} = \left(\frac{k_2 \sinh \xi_1 \sinh \xi_2 - k_1 \cosh \xi_1 \cosh \xi_2}{k_1 \sinh \xi_1 \sinh \xi_2 - k_2 \cosh \xi_1 \cosh \xi_2}\right)^2.$$

The same ψ_1 and $\psi_2 = \sinh \xi_2$ yield another potential with a power $4/5$ singularity:

$$u^{(2)} = \left(\frac{k_2 \sinh \xi_1 \cosh \xi_2 - k_1 \sinh \xi_2 \cosh \xi_1}{k_1 \sinh \xi_1 \cosh \xi_2 - k_2 \sinh \xi_2 \cosh \xi_1}\right)^2.$$

By construction, these potentials have only two levels $\lambda_{1,2}$.

[2]Note that the expressions for $u^{(2)}$ are parametric. In order to interpret the formulae correctly, the reader is referred back to (10.1), (10.6), and (10.7). The orders of the singularities pertain to the potential $\epsilon(x)$, which is a zero of the function $u(x)$ and a singularity of the potential $U(y) = v_{yy}/v_y$, where the function $v(y)$ solves the equation $u[v(y)] = v_y(y)$.

All the *regular* reflexionless potentials can be also built by the formulae (10.12) and (10.13) once again by dressing $U = 0$. Matveev and Salle [84] find super-reflexionless potentials for the KdV equation, alias positons. In the same vein one can operate on equations (10.1) and (10.7). In order to do so, one should use the formulae (10.12) and (10.13) with $n = 2$ choosing the prop solutions $\psi_{1,2}$ respectively as $\psi_1(y, \lambda_1)$ and $\psi_1(y, \lambda_1 + \delta)$, and then letting $\delta \to 0$. If with $U = 0$ and ψ_1 generates a single soliton potential, then (10.16) defines a single positon potential. In addition to B-potentials, various other interesting ones can be produced. For instance, one can construct soluble potentials with a finite equidistant spectrum[3]. One can also investigate potentials which change in a specific simple way under the Darboux transform, e.g., such that $U \to U+$ const. or $U \to$ const. $\times U$. For the Schrödinger equation, the former transformation is shape-invariant and results in the harmonic oscillator potential.

Example 10.2 (shape-invariant potential). Let

$$U^{(1)} = U + \frac{2}{\omega^2},$$

for a constant ω. Then we can obtain parametrically the function $u(x)$ from (10.1) as follows:

$$u(x) = \alpha z \exp\left(-\omega^2 z - \frac{\kappa^2}{z}\right), \qquad x = x_0 - \alpha\omega^2 \int dz \exp\left(-\omega^2 z - \frac{\kappa^2}{z}\right),$$

where, κ, x_0, α are real constants and $z = \exp(-y/\omega^2)$. The prop function ψ_1 rendering the potential $U^{(1)}$ from U has the countenance $\psi_1 = \exp(-\omega^2 z)$ and solves (10.7) with an eigenvalue $\lambda_1 = b^2/\omega^2$. It's easy to verify that the dielectric permittivity $\epsilon(x) = 1/u^2(x)$ has a second order pole at $x = x_0$.

The theory of Darboux transformations for the Schrödinger equation utilizes the concept of dressing chains of discrete symmetries and their closing. The work of Veselov and Shabat [102] elucidates how the dressing chain closing method can be used in order to obtain various potentials with meaningful mathematical physics. Namely, a simple closing procedure leads one to the harmonic oscillator potential (resulting also in a shape-invariant change of potential). A more complicated closing scheme results in finite-gap potentials as well as the fourth and the fifth Painleve equations, see [102].

Dressing chains can be written out for equation (10.7) as well. Let us introduce a sequence $\{f_n\}_{n\geq 1}$ of functions as follows:

$$f_n = D \ln \psi_n^{(n-1)},$$

with the quantity $\psi_n^{(n-1)}$ as it has been introduced in the diagram (10.15) above. In particular, it corresponds to the pre-chosen value λ_n of the spectral parameter.

[3]The same statement applies to the (stationary) Schrödinger equation.

One can verify, starting from $n = 1$, that

$$U^{(n)} = U - 2D \ln \prod_{j=1}^{n} f_j.$$

Besides, direct substitution shows that f_n satisfies the equation

$$f_n' + f_n^2 - U^{(n)} f_n = \lambda_n,$$

where $f' = Df$. The two latter relations imply the recursion connecting f_n and f_{n+1} as follows:

$$(f_n f_{n+1})' = f_n f_{n+1}(f_n - f_{n+1}) + \lambda_{n+1} f_n - \lambda_n f_{n+1}. \tag{10.18}$$

Equation (10.18) represents a dressing chain for the acoustic problem.

In a way analogous to the theory of dressing chains for the Schrödinger equation [102], we are interested in T-periodic chain closing, namely imposing the condition $f_{n+T} = f_n$ for an integer $T \geq 1$. We shall consider here the easiest case $T = 1$.

Example 10.3 (regular potential)**.** Given the spectral parameter values $\lambda_{1,2}$, one obtains a one-parameter family of potentials, indexed by a constant c:

$$U = \frac{-(\lambda_1 - \lambda_2)^2 y^2 + 2c(\lambda_2 - \lambda_1)y + 6\lambda_1 - 2\lambda_2 - c^2}{2[(\lambda_1 - \lambda_2)y + c]}.$$

If $\lambda_2 = 3\lambda_1 > 0$ and $c = 0$, we can express the function $u(x)$ parametrically:

$$x(y) = \frac{\sqrt{\pi}}{2\alpha} Erf(\alpha y), \quad u(y) = \exp\left(-\alpha^2 y\right),$$

where $\alpha^2 = -\lambda/2 > 0$.

It is known that for the Schrödinger equation, a nontrivial chain closing operation with $T > 1$ results in finite gap potentials [102]. Such potentials for the HD equation are due to Dmitrieva [32]. A close analogy between the Schrödinger equation and the acoustic problem (see [24, 31, 44]) suggests that one can expect results similar to those of [102] apropos of the analysis of higher order chain closing for $T > 1$. We expect that potentials built in such a way can have interesting physical applications, such as for instance a model of wave propagation in media whose dielectric permittivity is a periodic function of a single spatial variable.

10.5 Harry-Dym Equation

The Harry-Dym (HD) equation

$$u_t = u^3 u_{xxx} + \beta u_x, \tag{10.19}$$

with some real constant β, has been studied quite extensively since late 70's, see e.g., [31, 32, 44] and references therein. The principal approach to it has been based on its relation to the KdV, mKdV and other more classical hierarchies of integrable PDEs [31, 32]. However, as mentioned before, this relation is not straightforward, and the direct approach developed herein in principle enables one to produce a wider range of solutions of the HD equation.

The acoustic problem (10.1) is the first equation in the Lax pair for the HD equation (10.19), the full pair is

$$\psi_{xx} = \frac{\lambda}{u^2}\psi, \quad \psi_t = (4\lambda u + \beta)\,\psi_x - 2\lambda u_x \psi. \tag{10.20}$$

The coordinate change (10.6) in the presence of time dependence becomes

$$t \to t, \quad x \to v(y,t),$$

thus

$$\partial_x \to \frac{1}{v_y}\partial_y, \quad \partial_t \to \partial_t - \frac{v_t}{v_y}\partial_y.$$

After this change, (10.20) becomes

$$\psi_{yy} = \frac{v_{yy}}{v_y}\psi_y + \lambda\psi, \quad \psi_t = \left(\frac{v_t+\beta}{v_y} + 4\lambda\right)\psi_y - \frac{2\lambda v_{yy}}{v_y}\psi, \tag{10.21}$$

and the HD equation (10.19) is transformed into

$$v_y(v_t v_{yy} - v_{yt}v_y) + 3v_{yy}^2 + v_y(v_{4y}v_y - 4v_{3y}v_{yy} + \beta v_{yy}) = 0. \tag{10.22}$$

with the notations v_{3y}, v_{4y} for the partial derivatives in y of order 3 and 4 respectively.

The goal now is to extend the Darboux transformation (10.12) for equation (10.7) alias the first equation in (10.21), so that it agrees with the second equation in the Lax pair. At this point, it only provides the value of the partial derivative

$$v_y^{(1)} = \left(\frac{\psi_1}{\psi_{1,y}}\right)^2 \equiv A(y,t),$$

rather than the dressed quantity $v^{(1)}$ of interest.

Lemma 10.4. *The Darboux transformation is an S-symmetry for the Lax pair* (10.21), *and therefore for the HD equation* (10.22).

Proof: Let $v_t^{(1)} = B(y,t)$ be an unknown and let's assume that the dressed function $\psi^{(1)}$ according to (10.12) satisfies the second equation of the pair (10.21) with $v^{(1)}$, and the quantities (λ, β) remain the same. One can express the unknown quantity B as follows:

$$B = \left(\frac{\psi_1}{\psi_{1,y}}\right)^2 (\beta + 4\mu v_y + v_t - 2v_y U_y) + \frac{4\psi_1 v_{yy}}{\psi_{1,y}} - 4v_y - \beta,$$

and verify that $B_y = A_t$. It follows that

$$v^{(1)}(y,t) = \int A dy + B dt, \tag{10.23}$$

with a closed 1-form under the integral. □

Example 10.5. (single soliton solution and positon solution): Let $v = y$, $\lambda_1 = k^2$, $\psi_1(y,t,k) = \sinh[\phi(y,t,k)]$ with $\phi = k[y + (4k^2 + \beta)t]$, then by (10.23) one has

$$v^{(1)} = \frac{1}{k^3}(\phi - \tanh\phi) - (4+\beta)t.$$

The function $v^{(1)}(y,t,k)$ determines a single soliton B-potential $U^{(1)} = v^{(1)}_{yy}/v^{(1)}_y$. The single positon potential is obtained from two distinct soliton solutions $\psi_1(y,t,k)$ and $\psi_1(y,t,k+\delta)$, using them as the prop functions $\psi_{1,2}$ in the formulae (10.12) and (10.13) with $n = 2$ and taking the limit as $\delta \to 0$. Namely,

$$v^{(2)}_y = v_y \left(\frac{\psi_{1,yyk}\psi_{1,y} - \psi_{1,yy}\psi_{1,yk}}{\psi_{1,yk}\psi_1 - \psi_{1,y}\psi_{1,k}} \right)^2,$$

where the subscript $_k$ means differentiation in k. Taking ψ_1 explicitly as the hyperbolic sine in the previous example results in

$$v^{(2)}_y = k^4 \left(\frac{\sinh 2\phi + 2\tilde{\phi}}{\sinh 2\phi - 2\tilde{\phi}} \right)^2,$$

with $\tilde{\phi} = k[y + (12k^2 + \beta)t]$.

10.6 Modified Harry-Dym Equation

It is well known that the dressing formalism enables one to produce hierarchies of integrable PDEs. Borisov and Zykov [22] proposed a technique for proliferation of integrable equations, which they applied to the KdV and the Sine-Gordon (SG) equations. The technique is based on dressing chain closing. The main idea of the approach is as follows: Take KdV equation for example, rewrite it as a compatibility condition of a pair of equations denoted by L_1 and A_1. Using invariance of the pair with respect to the Darboux transformation (which is viewed as a discrete symmetry), a second pair L_2 and A_2 is built. Eliminating the potentials from L_1 and L_2, and A_1 and A_2, it is possible to obtain two equations called an x and a t chains [22] denoted by C_x and C_t. If a potential is eliminated from L_1 and A_1, one ends up with a modified KdV. This procedure can be repeated, producing new equations with their Lax pairs, e.g., m^{2}KdV, m^{3}KdV etc. The former becomes the exponential Calogero-Degasperis equation [24] after an exponential change, and the latter contains an elliptic equation of the same authors.

Using the technique described, it was shown [88] that the m^{N}KdV equations with $N = 0, \ldots, 3$ together with the Krichever-Novikov equation, exhaust (modulo a contact transformation) all the integrable equations of the form $u_t + u_{xxx} + f(u_{xx}, u_x, u) = 0$. Using the same approach to the sine-Gordon equation, Borisov and Zykov [22] obtained a new nonlinear equation in the second step. This equation has a non-trivial Bäcklund transform, admitting an interesting 2π-kink-shelf solution. The same technique was also shown to be applicable to the study of more difficult (1+2)-dimensional nonlinear PDEs. For instance, in [109], the proliferation procedure was successfully adapted to the Kadomtsev-Petviashvili and Boiti-Leon-Pempinelli equations.

Let us apply this formalism to the HD equation. First note that the Lax pair for (10.22) can be written as a system of two Ricatti equations:

$$\begin{aligned} g_y &= -\lambda g^2 - Ug + 1, \\ g_t &= \lambda\left(2U_y - \frac{v_t+\beta}{v_y} - 4\lambda\right)g^2 - \left(\frac{v_{ty}}{v_y} + 4\lambda U\right)g + \frac{v_t+\beta}{v_y} + 4\lambda, \end{aligned} \tag{10.24}$$

where the function $g = g(y,t)$ is connected with the solution ψ of (10.21) as $g = \psi_y/\psi$. Eliminating the v from (10.24) and returning to the old variables via $x = g$, $u = g_y$, we obtain a modified Harry Dym (mHD) equation:

$$u_t = u^3 u_{3x} + 3u^2 u_x u_{xx} - 3\lambda^2 x u^2 - \frac{3u^2(uu_{xx} + u_x^2)}{x} + \frac{6u^3 u_x}{x^2} + \frac{3u^2(1-u^2)}{x^3}. \tag{10.25}$$

Note that (10.25) can be rewritten quite nicely in the new variables $x = 1/z$, $u(x,t) = \sqrt{\theta(z,t)}$:

$$\left(\theta^{-1/2}\right)_t = \frac{3\lambda^2}{z} + z^3\left(\frac{1}{2}z^3\theta_{3z} + \frac{9}{2}z^2\theta_{zz} + 9z\theta_z + 3\theta - 3\right), \tag{10.26}$$

which can be further simplified as

$$\left[\left(e^{-2\xi}\eta + 1\right)^{-1/2}\right]_t = 3\lambda^2 e^{-\xi} + \frac{1}{2}e^{\xi}\left(\eta_{3\xi} - \eta_\xi\right),$$

by another change of variables

$$\theta = e^{-2\xi}\eta(\xi,t) + 1, \quad \xi = \ln z.$$

The dressing chain method produces not only equation (10.26), but also its Lax pair. It is constructed as follows: Return to the chain (10.18) and let $f_n = 1/g_n$, $f_{n+1} = \Psi$, $\lambda_n = \lambda$, $\lambda_{n+1} = \mu$. Regarding μ as a spectral parameter, one can see that (10.18) can be viewed as an L-equation of the Lax pair for equation (10.26). One can also define the second non-stationary chain C_t for the functions g_n in terms of the temporal equation in (10.21) and build the A-equation. Omitting the lengthy but straightforward computation, we present the Lax pair for equation (10.26), written in the variables t, z:

$$\Psi_z = \frac{\mu}{z^2\sqrt{\theta}}\Psi^2 + \left(\frac{1}{z} + \frac{1}{z\sqrt{\theta}} - \frac{\mu}{z^3\sqrt{\theta}}\right)\Psi - \frac{1}{z^2\sqrt{\theta}}, \qquad (10.27)$$
$$\Psi_t = \mu a\Psi^2 + b\Psi + c,$$

where

$$a = -4\mu + 2\lambda - \frac{\lambda^2}{z^2} - 2\lambda\sqrt{\theta} + (\theta - 1)\, z^2 + 2z^3\theta_z + \frac{1}{2}z^4\theta_{zz},$$
$$b = 4\left(\frac{\lambda}{z} - z - z\sqrt{\theta}\right)\mu + \frac{1}{2}z^5\theta_{zz} + 2z^4\theta_z + \left(\theta - \frac{\lambda}{2}\theta_{zz} - 1\right)z^3$$
$$-3\lambda z^2\theta_z + 3\lambda(1-\theta)z - \frac{3\lambda^2}{z} + \left(\frac{\lambda}{z}\right)^3,$$
$$c = 4\mu - \frac{1}{2}z^4\theta_{zz} - 3z^3\theta_z + \left(1 - 3\theta - 2\sqrt{\theta}\right)z^2 - 2\lambda + \left(\frac{\lambda}{z}\right)^2.$$

Note that the spectral parameter in (10.27) is μ, whereas λ enters the non-linear equation (10.26). As one can see, the Lax pair for (10.26) also has the form of a pair of Ricatti equations. These equations can be simultaneously linearized in order to represent (10.26) as a compatibility condition for two linear equations, as it is done in the theory of solitons.

We will not proceed further with the mHD equation. To conclude this section, we would like to repeat the statement that the exact solutions of (10.26) can be easily found via the dressing technique, and this procedure can be extended in order to produce the mHD equation and its Lax pair. It is worthy of emphasizing that HD and (10.26) are members of different hierarchies. Discrete symmetries enable one to establish connections between different integrable hierarchies in a way apparently more systematic than trying to guess the Miura transformation.

10.7 Moutard Transformations

In this section, we return to the PDE (10.5) which was obtained from the Maxwell equations (10.2) in the case of an isotropic but inhomogeneous two-dimensional (x, y) medium. One can obtain the counterpart of Darboux transformation for the PDE (10.5), known also as the Moutard transformation [84].

Lemma 10.6. *Let $\psi = \psi(x, y)$ and $\phi = \phi(x, y)$ be two particular solutions of (10.5), i.e.,*

$$\Delta\psi - \lambda\epsilon\psi = \Delta\phi - \lambda\epsilon\phi = 0. \qquad (10.28)$$

The following transformation is the Moutard transformation:

$$\psi \to \psi^{(1)} = \frac{\theta[\psi, \phi]}{\phi}, \qquad \epsilon \to \epsilon^{(1)} = \epsilon - 2\lambda\Delta\ln\phi, \qquad (10.29)$$

where

$$\theta[\psi,\phi] = \int_\Gamma dx_\mu \varepsilon_{\mu\nu} \left(\phi\partial_\nu\psi - \psi\partial_\nu\phi\right), \tag{10.30}$$

with the following tensor notations: $\mu \in \{1,2\}$, $x_\mu \in \{x,y\}$, $\partial_\mu = \partial/\partial x_\mu$, $\varepsilon_{\mu\nu}$ *is a fully antisymmetric tensor with* $\varepsilon_{12} = 1$, *summation is implied over repeated indices.*

The Moutard transformation (10.29) can be iterated several times, and the result can be expressed via Pfaffian forms [84]. In terms of the original variables of the Maxwell equations (10.2). A straightforward computation (recall, $\lambda = -\frac{c^2}{\omega^2}$) yields the expressions for the dressed electric and magnetic fields $\mathbf{E}^{(1)}$, $\mathbf{B}^{(1)}$:

$$\mathbf{E}^{(1)} = e^{i\omega t}(0,0,\psi^{(1)}), \quad \mathbf{B}^{(1)} = \frac{c}{\omega}e^{i\omega t}(-\psi^{(1)}_y, \psi^{(1)}_x, 0), \quad \mathbf{D}^{(1)} = \epsilon^{(1)}\mathbf{E}^{(1)}. \tag{10.31}$$

On the basis of (10.31) one can build a variety of exact solutions of the Maxwell equations.

Example 10.7. (singular potentials): As a simple example, let's dress $\epsilon = 0$. This isn't quite a medium, but one can easily proceed with formal calculations (10.29) and (10.31) which result in a new "medium" whose dielectric permittivity $\epsilon^{(1)}(x,y)$ and the stationary component of the field $\psi^{(1)}$ are as follows:

$$\epsilon^{(1)} = -\frac{8c^2}{\omega^2}\frac{a'(z)b'(\bar z)}{\left(a(z)+b(\bar z)\right)^2}, \quad \psi^{(1)} = \frac{a(z)\beta(\bar z) - \alpha(z)b(\bar z) + \xi(z,\bar z)}{a(z)+b(\bar z)}, \tag{10.32}$$

where

$$\xi(z,\bar z) = \int dz\left(\alpha(z)a'(z) - a(z)\alpha'(z)\right) + \int d\bar z\left(\beta'(\bar z)b(\bar z) - b'(\bar z)\beta(\bar z)\right),$$

$a(z)$, $\alpha(z)$, $b(\bar z)$, $\beta(\bar z)$ are arbitrary functions of $z = x+iy$, $\bar z = x - iy$. Note that the function $\psi^{(1)}$ from (10.29) and (10.31) provides in fact a general solution of the dressed equation, for it is described in terms of two arbitrary functions $\alpha(z)$ and $\beta(\bar z)$. To ensure that the quantities found correspond to a physical non-absorbing medium, one should require that the dressed dielectric permittivity function $\epsilon^{(1)}$ to be real. This imposes an extra restriction to the quantities $a(z)$ and $b(\bar z)$, namely $b(\bar z) = \overline{a(z)}$. Generally speaking, the functions $\epsilon^{(1)}$ and $\psi^{(1)}$ from (10.32) will have singularities along certain curves in the (x,y)-plane.

The reflexionless B-potentials for the one-dimensional problem (10.1) above (see Example 10.1) possess point singularities on the real line (corresponding to zeroes of the function $u(x)$). Clearly, their 2D-analogues, such as (10.32) for equation (10.5) allow a much more diverse structure of singularities on the real plane. On the other hand, not requiring that the quantity $\epsilon^{(1)}$ be real, one obtains an absorbing medium which may not be devoid of interest for physical applications.

Finally, let us study a dressing chain generated by the Moutard transformations (10.29). A simple periodic closing of the dressing chain results in a regular dielectric permittivity, similar to the 1D case studied above. Denote by $f_n = \ln \phi$, $f_{n+1} = \ln \psi^{(1)}$. Then after a straightforward computation

$$\Delta (f_n + f_{n+1}) = \|\nabla f_n\|^2 - \|\nabla f_{n+1}\|^2, \tag{10.33}$$

where $\|\cdot\|$ is the Euclidean norm. The chain (10.33) is closely related to that of Veselov and Shabat [102] for the Schrödinger equation. Choosing f_n specifically as

$$f_n = \sqrt{\lambda_n} y + \int dx\, g_n(x),$$

and substituting it into (10.33) (λ_n being constant), we obtain for the quantities $g_n(x)$ the following expression

$$(g_n + g_{n+1})' = g_n^2 - g_{n+1}^2 + \lambda_n - \lambda_{n+1},$$

matching the corresponding formula of [102].

Example 10.8. (regular dielectric permittivity): The simplest periodic closing of the dressing chain (10.33) is $f_{n+1} = f_n = F(x, y)$, which implies that the latter function F is harmonic, and that the regular dielectric permittivity function in the corresponding medium is given by the formula

$$\epsilon(x, y) = \frac{c^2}{\omega^2} \left(F_x^2 + F_y^2\right).$$

Chapter 11

SUSY and Spectrum Reconstructions

Supersymmetric quantum mechanics realizes the quantum description of systems with double degeneracy of energy levels. In one dimension, the supersymmetry is intrinsically connected with the Darboux transformation (DT) [29, 47]. The DT connects two Hamiltonians h_0 and h_1 with equivalent spectra:

$$h_0 = q^+ q + E_0, \quad h_1 = q q^+ + E_0, \quad q \equiv \frac{d}{dx} - (\ln \varphi)', \tag{11.1}$$

where q^+ is the Hermitian conjugate of the q operator, and φ is a solution to the equation $h_0 \varphi = E_0 \varphi$. It is easy to see that h_0 and h_1 are intertwined by q and q^+

$$q\, h_0 = h_1\, q, \quad h_0\, q^+ = q^+\, h_1, \tag{11.2}$$

and therefore

$$\psi^{(1)} = q\, \psi, \quad \psi = q^+\, \psi^{(1)}, \tag{11.3}$$

if

$$h_0 \psi = E_1 \psi, \quad h_1 \psi^{(1)} = E_1 \psi^{(1)}, \tag{11.4}$$

and $\varphi^{(1)} = \varphi^{-1}$ solves $h_1 \varphi^{(1)} = E_0 \varphi^{(1)}$. For brevity, the normalizing multipliers in (11.3) are omitted.

DT allows us to construct one-dimensional potentials with arbitrary preassigned discrete spectrum. For instance, if the support function $\varphi(E_0; x)$ is a wave function (w.f.) of the ground state h_0, then the discrete spectrum of h_1 coincides with the spectrum of h_0 without lower level E_0 [28]. It is possible to add the level E_0 (missing in the spectrum of h_0) to the spectrum of h_1. To achieve this, it is sufficient to exploit such φ that

$$\varphi \to +\infty, \quad x \to \pm\infty,$$

and φ is a positive definite function of x. It is convenient to choose φ as:

$$\varphi = \lambda \varphi_+ + (1 - \lambda) \varphi_-, \tag{11.5}$$

where φ_+ and φ_- are positive definite functions with the following asymptotic behavior:

$$\varphi_\pm \to \begin{cases} +\infty, & \text{for } x \to \pm\infty \\ 0, & \text{for } x \to \mp\infty, \end{cases} \tag{11.6}$$

and λ is a real parameter in the interval $[0, 1]$. If $0 < \lambda < 1$, then the level E_0 is on the lower boundary of the spectrum of h_1. If $\lambda = 0$ or $\lambda = 1$, the level E_0 is not in the spectra of h_0 and h_1 and the spectra are the same.

This relates to the one-dimensional supersymmetric quantum mechanics based on the following commutation relations:

$$[Q, H] = [Q^+, H] = 0, \quad \{Q, Q^+\} = H, \tag{11.7}$$

where

$$Q = q\sigma_+, \quad Q^+ = q^+\sigma_-, \quad H = \text{ diag } (h_0 - E_0, h_1 - E_0),$$

and $\sigma_\pm = (\sigma_1 \pm i\sigma_2)/2$, and $\sigma_{1,2}$ are Pauli matrices. Notice that q and q^+ are bosonic, and σ_- and σ_+ are fermionic, operators of creation-annihilation. If the spectra of h_0 and h_1 differ by the only level, then (11.7) corresponds to the exact supersymmetry. If the level E_0 is absent in the spectra of both operators, then the supersymmetry is broken. It is easy to see that the level E_0 cannot simultaneously appear in the spectra of h_0 and h_1.

11.1 SUSY in Two Dimensions

In this section, we consider the case of two-dimensional supersymmetric quantum mechanics. The explicit form of operators that satisfy the algebraic relation (11.7) in 2D is determined by the expressions:

$$Q = \begin{pmatrix} 0 & 0 & 0 & 0 \\ q_1 & 0 & 0 & 0 \\ q_2 & 0 & 0 & 0 \\ 0 & q_2 & -q_1 & 0 \end{pmatrix}, \quad Q^+ = \begin{pmatrix} 0 & q_1^+ & q_2^+ & 0 \\ 0 & 0 & 0 & q_2^+ \\ 0 & 0 & 0 & -q_1^+ \\ 0 & 0 & 0 & 0 \end{pmatrix}, \tag{11.8}$$

$$H = \text{ diag } (h_0 - E_0, \tilde{h}_{lm} - 2\delta_{lm}E_0, h_1 - E_0), \tag{11.9}$$

where

$$h_0 = q_m^+ q_m + E_0, \quad h_1 = q_m q_m^+ + E_0, \quad \tilde{h}_{lm} \equiv h_{lm} + H_{lm} - E_0\delta_{lm}, \tag{11.10}$$

and

$$h_{lm} = q_l q_m^+ + E_0\delta_{lm}, \quad H_{lm} = p_l p_m^+ + E_0\delta_{lm}, \tag{11.11}$$

and $q_l = \partial_l - \partial_l(\ln\varphi)$, $p_l = \varepsilon_{lk}q_k^+$, where ε_{lk} is the antisymmetric tensor, $\partial_l \equiv \partial/\partial x^l$, $(l = 1, 2)$, and the sum by repeated indices is implied.

In contrast to 1D, there is no association between the spectra of h_0 and h_1 here. In 2D, there is no formula expressing wave functions of h_1 via wave functions of h_0. The existence of such a formula implies the connection between the spectra of h_0 and h_1. However, for special potentials, such a connection can still exist. Moreover, there could be expressions that connect wave functions of h_0 and h_1, that are not related to a physical spectrum.

The general coupling between the spectra exists for h_0 and h_{lm} or h_1 and H_{lm}. Notice that h_1 may be represented as $h_1 = p_m^+ p_m + E_0$, it is easy to verify the relations:

$$q_l h_0 = h_{lm} q_m, \quad p_l h_1 = H_{lm} p_m, \quad h_0 q_l^+ = q_m^+ h_{ml}, \quad h_1 p_l^+ = p_m^+ H_{ml}. \tag{11.12}$$

The same formulae apply to the operator, $\tilde{h}_{lm}$, h_0, and h_1, perhaps without the level E_0.

The supersymmetry defined by the operators (11.8) and (11.9), was studied in [7, 10, 28], with the assumption that φ is a wave function of the basic state of the Hamiltonian h_0. It was shown that such a choice of φ leads to the assertion that the level E_0 is either absent from the physical spectra of $\tilde{h}_{lm}$ and h_1, or related to the unbroken supersymmetry. In the following sections, we study the inverse problem, i.e., addition of the level E_0, that is absent from the spectrum of h_0, to the spectra of $\tilde{h}_{lm}$ and h_1.

11.2 The Level Addition

Let $u = u(x, y)$ be an integrable potential, i.e., one can solve the Schrödinger equation $h_0\psi = E\psi$ explicitly for any spectral parameter value E, $h_0 = -\Delta + u$. Unlike the one-dimensional case, the potential

$$u^{(1)} = u - 2\Delta \ln \varphi, \tag{11.13}$$

where φ is a support function, is not integrable here. Assume that the spectral parameter value E_0 lies below the basic state energy of the Hamiltonian h_0, how should one choose the support function φ such that the level E_0 appears in the physical spectra of h_1 and $\tilde{h}_{lm}$?

For the scalar Hamiltonian h_1, it is easy to verify that the function φ^{-1} satisfies the equation

$$h_1 \frac{1}{\varphi} = E_0 \frac{1}{\varphi}, \tag{11.14}$$

therefore, it is enough to choose φ as a positive function of x and y, which grows exponentially in all directions on the plane.

For the matrix Hamiltonian $\tilde{h}_{lm}$, if the function ψ is a solution of the Schrödinger equation with E_0, then the function

$$\tilde{\psi}_m = q_m \psi, \tag{11.15}$$

satisfies the equation

$$\tilde{h}_{lm}\tilde{\psi} = E_0\tilde{\psi}. \tag{11.16}$$

One can show that the level E_0 belongs to the spectrum of $\tilde{h}_{lm}$, iff $\tilde{\psi}_m$ is normable and satisfies the condition:

$$h_{lm}\tilde{\psi}_m = H_{lm}\tilde{\psi}_m = E_0\tilde{\psi}_m. \tag{11.17}$$

In fact, let $\tilde{\psi}_m$ be such that

$$\tilde{h}_{lm}\tilde{\psi}_m = E_0\tilde{\psi}_m, \quad (\tilde{\psi}_m, \tilde{\psi}_m) = 1. \tag{11.18}$$

Define the functions ρ_m and σ_m by

$$\rho_m \equiv h_{lm}\tilde{\psi}_m, \quad \sigma_m \equiv H_{lm}\tilde{\psi}_m. \tag{11.19}$$

From (11.10) and (11.11), it follows that $\sigma + \rho = 2E_0\tilde{\psi}$ (indices omitted), i.e.,

$$(\rho + \sigma, \rho + \sigma) = 4E_0^2, \tag{11.20}$$

if ρ and σ are normable. Otherwise one may check that

$$h_{mk}H_{kl} = H_{mk}h_{kl} = E_0\tilde{h}_{ml}. \tag{11.21}$$

From (11.21), it follows that

$$h_{lm}\sigma_m = H_{lm}\rho_m = E_0^2\tilde{\psi}_l. \tag{11.22}$$

Hence

$$(\tilde{\psi}_m, h_{lm}\sigma_l) = (h_{lm}\tilde{\psi}_m, \sigma_l) = (\rho_m, \sigma_m) = E_0^2. \tag{11.23}$$

Combining with (11.20), we obtain $(\rho - \sigma, \rho - \sigma) = 0$, therefore $\sigma_m = \rho_m = E_0\tilde{\psi}_m$. Finally from (11.19), we get (11.17). Thus for the level E_0 lying in the physical spectrum of $\tilde{h}_{lm}$, it is necessary to find a normable solution of (11.17). Let $\tilde{\psi}_m$ be such a solution, applying $\tilde{h}_{lm}$, one gets

$$q_m^+\tilde{\psi}_m = p_m^+\tilde{\psi}_m = 0. \tag{11.24}$$

This means that there exist two functions ψ and $\psi^{(1)}$ such that

$$\tilde{\psi}_m = q_m\psi = p_m\psi^{(1)}, \tag{11.25}$$

where

$$h_0\psi = E_0\psi, \quad h_1\psi^{(1)} = E_0\psi^{(1)}. \tag{11.26}$$

Solving (11.25) with respect to $\psi^{(1)}$, we get the important relation:

$$\psi^{(1)} = \frac{1}{\varphi}\int dx_k\varepsilon_{km}(\varphi\partial_m\psi - \psi\partial_m\varphi), \tag{11.27}$$

that is known as a Moutard transformation [16]. From the relations among $\tilde{\psi}_m$, ρ_m, and σ_m, the formula (11.15) obviously follows.

11.3 Potentials with Cylindrical Symmetry

Let ψ and $\varphi > 0$ be the solutions described at the end of the previous section. For the construction of matrix potentials with the level E_0, it is convenient to introduce an auxiliary function

$$f \equiv \frac{\psi}{\varphi} \tag{11.28}$$

that satisfies the equation

$$\partial_m(\varphi^2 \partial_m f) = 0. \tag{11.29}$$

Then

$$\tilde{\psi}_m = \varphi \partial_m f. \tag{11.30}$$

Consider the case when the seed potential possess the cylindrical symmetry $u = u(r)$. Integrating (11.29) and substituting in (11.30), one gets

$$\tilde{\psi}_m = \frac{x_m}{r^2 \varphi}. \tag{11.31}$$

The normalizing integral of (11.31) converges if φ grows at infinity as a power function, and at the vicinity of zero, it behaves as r^{-k}, $k > 0$. If one requires that the asymptotic behavior of φ is determined by the conditions:

$$\varphi \to \begin{cases} r^a, & \text{for } x^2 + y^2 \to \infty \\ r^b, & \text{for } x^2 + y^2 \to 0, \end{cases} \tag{11.32}$$

where $a > 1$, $b < 1$, then the normalizing integral of φ^{-1} should converge as well. This means that the level E_0 will be present in both spectra of the operators h_1 and $\tilde{h}_{lm}$ simultaneously.

The spectrum of the supersymmetrical Hamiltonian (11.9) consists of the levels

$$\{E_i - E_0, E_i^{(1)} - E_0\},$$

where E_i, $E_i^{(1)}$ are the levels of the discrete spectra of h_0 and h_1, respectively. If the condition (11.32) is satisfied, then in the spectrum of (11.9), there is a doubly degenerated level $E = 0$ to which the following eigenfunctions correspond,

$$\Psi_1 = \begin{pmatrix} 0 \\ 0 \\ 0 \\ \frac{1}{\varphi} \end{pmatrix}, \quad \Psi_2 = \begin{pmatrix} 0 \\ \varphi \partial_1 f \\ \varphi \partial_2 f \\ 0 \end{pmatrix}. \tag{11.33}$$

Using the explicit form of the odd supersymmetric operators (11.8), one can prove the relations for wave functions of the zero level

$$Q\Psi_{1,2} = Q^+\Psi_{1,2} = 0.$$

As an example, choose $\varphi = \exp(br)/r^k$, where $b, k > 0$. It satisfies the necessary asymptotic (11.32). As a result, we obtain two scalar potentials of the Hamiltonians h_0 and h_1:

$$u = \frac{k^2}{r^2} - \frac{b(2k-1)}{r}, \quad u^{(1)} = \frac{k^2}{r^2} - \frac{b(2k+1)}{r}. \tag{11.34}$$

The added level corresponds to the energy $E_0 = -b^2$. The discrete spectrum is determined by the formula:

$$E_N = -\frac{b^2(2k \mp 1)^2}{[1 + 2(N + \sqrt{m^2 + k^2})]^2}, \tag{11.35}$$

where the sign "−" corresponds to u, and "+" to $u^{(1)}$, and N is the principal and m – magnetic quantum numbers. The constructed potentials exhibit the difference between DT in multidimensions and one dimension. Specifically, the comparison of the spectra of the Hamiltonians h_0 and h_1 shows that the addition of the lowest level shifts all the spectrum. For the potential (11.34), it can be shown that when

$$k = \frac{(N+1)^2 - m^2}{2(N+1)},$$

the addition of the level $E_0 = -b^2$, does not move the exited level with the number N for fixed m. In general, the levels of the Hamiltonian h_1 go down with respect to the levels of h_0. The maximal displacement happens in the lowest part of the spectrum. The spectrum of the supersymmetric Hamiltonian (11.9) is doubly degenerated at the level $E = 0$. Its normable vacuum wave functions are given by the expressions (11.33), and

$$\frac{1}{\varphi} = r^k \exp(-br), \quad \partial_m f = \frac{x_m}{(r\varphi)^2}. \tag{11.36}$$

11.4 Extended Supersymmetry

In the previous section, we had demonstrated how to build a two-dimensional Hamiltonian with doubly degenerated level $E = 0$. One may obtain models with all degenerated levels with extended symmetry [9, 92]:

$$\{Q_i, Q_k\} = \delta_{ik} H, \quad [Q_i, H] = 0, \quad i, k = 1, \ldots, N. \tag{11.37}$$

First one can define two Hamiltonians: H_1 is determined by (11.9), and H_2 differs from it by the permutation of h_0 and h_1. Then one can define three operators: Q_1 is defined by (11.8), and

$$Q_2 = \begin{pmatrix} 0 & 0 & 0 & 0 \\ -q_2^+ & 0 & 0 & 0 \\ q_1^+ & 0 & 0 & 0 \\ 0 & -q_1^+ & -q_2^+ & 0 \end{pmatrix}, \quad B = \begin{pmatrix} 0 & q_2 & -q_1 & 0 \\ -q_1 & 0 & 0 & q_2^+ \\ -q_2 & 0 & 0 & -q_1^+ \\ 0 & -q_1^+ & -q_2^+ & 0 \end{pmatrix}, \tag{11.38}$$

that satisfy the commutation relations

$$Q_2^+B + BQ_1^+ = Q_2B + BQ_1 = BH_1 - H_2B = 0, \tag{11.39}$$
$$H_1 = \{Q_1^+, Q_1\} = B^+B, \quad H_2 = \{Q_2^+, Q_2\} = BB^+. \tag{11.40}$$

One can verify that

$$H(2) = \mathrm{diag}(H_1, H_2), \quad Q_1(2) = \mathrm{diag}(Q_1, Q_2), \tag{11.41}$$
$$Q_2(2) = \begin{pmatrix} 0 & 0 \\ B & 0 \end{pmatrix}, \tag{11.42}$$

form the algebra (11.37), when $N = 2$. The introduced operators are the building blocks for the construction of the extended supersymmetry matrices for any N. It is possible to demonstrate that at every step there is an operator factorizing a superhamiltonian as in (11.40). At the step N, the superhamiltonian

$$H(N) = \begin{pmatrix} B_N^+B_N & 0 \\ 0 & B_NB_N^+ \end{pmatrix}, \tag{11.43}$$

is factorized by the operator $B_{N+1} = \mathrm{diag}(B_N, B_N^+)$. In turn, $H(N+1)$, $B_k(N+1)$, $(k \leq N)$ are defined by the substitution $B_N \to B_{N+1}$ and the addition of the new operator $Q_{N+1}(N+1)$ in the new matrix. At every step the dimension of matrices duplicates, therefore, the corresponding algebra is realized by $2^{N+1} \times 2^{N+1}$ matrices. For example, at $N = 4$, there would be four operators Q_i of dimension 32×32 and the superhamiltonian H:

$$H \equiv H(4) = \mathrm{diag}(H_1, H_2, H_2, H_1, H_2, H_1, H_1, H_2), \tag{11.44}$$
$$Q_1(4) = \mathrm{diag}(Q_1, Q_2, Q_2, Q_1, Q_2, Q_1, Q_1, Q_2), \tag{11.45}$$
$$Q_s(4) = \begin{pmatrix} 0 & 0 & 0 & 0 & 0 & 0 & 0 & 0 \\ B & 0 & 0 & 0 & 0 & 0 & 0 & 0 \\ B & 0 & 0 & 0 & 0 & 0 & 0 & 0 \\ 0 & B^+ & -B^+ & 0 & 0 & 0 & 0 & 0 \\ B & 0 & 0 & 0 & 0 & 0 & 0 & 0 \\ 0 & B^+ & 0 & 0 & -B^+ & 0 & 0 & 0 \\ 0 & 0 & B^+ & 0 & -B^+ & 0 & 0 & 0 \\ 0 & 0 & 0 & B & 0 & -B & B & 0 \end{pmatrix}, \tag{11.46}$$

and $s = 2, 3, 4$. The nonzero elements of the three operators in (11.46) are under the diagonal. If as a basic scalar model, one takes the potential considered in the last section, it is obvious that all levels including zero, are degenerate with the multiplicity 2^N.

Chapter 12

Darboux Transformations for Dirac Equation

12.1 Dirac Equation

Let us consider the four-dimensional Dirac equation

$$(i\gamma^\mu \partial_\mu - \gamma^\mu A_\mu(t,x) + m)\Psi = 0. \tag{12.1}$$

We use the γ–matrix representation

$$\gamma^0 = \begin{pmatrix} 0 & \sigma_2 \\ \sigma_2 & 0 \end{pmatrix}, \quad \gamma^1 = \begin{pmatrix} 0 & -i\sigma_1 \\ -i\sigma_1 & 0 \end{pmatrix},$$

$$\gamma^2 = \begin{pmatrix} -iI & 0 \\ 0 & iI \end{pmatrix}, \quad \gamma^3 = \begin{pmatrix} 0 & -i\sigma_3 \\ -i\sigma_3 & 0 \end{pmatrix},$$

and for $A_\mu(t,x)$ we have either:

$$A_\mu(t,x) = (0, 0, A_2(t,x), A_3(t,x))^T, \tag{12.2}$$

or

$$A_\mu(t,x) = (0, 0, Q(t,x), cQ(t,x))^T, \quad c = \text{const.} \tag{12.3}$$

One can easily see that (12.2) and (12.3) are solutions of Maxwell equations. Let Ψ have the form:

$$\Psi = \Phi(t,x)\exp[i(py + qz)], \quad \text{Im } p = \text{Im } q = 0. \tag{12.4}$$

Then Φ satisfies the Zakharov–Shabat equation:

$$\Phi_t = J\Phi_x + U\Phi. \tag{12.5}$$

For the case (12.2):

$$U = \begin{pmatrix} 0 & \tilde{A} & 0 & m - \tilde{B} \\ \tilde{A} & 0 & \tilde{B} - m & 0 \\ 0 & \tilde{B} + m & 0 & \tilde{A} \\ -\tilde{B} - m & 0 & \tilde{A} & 0 \end{pmatrix}, \quad J = \begin{pmatrix} 1 & 0 & 0 & 0 \\ 0 & -1 & 0 & 0 \\ 0 & 0 & 1 & 0 \\ 0 & 0 & 0 & -1 \end{pmatrix},$$

where $\tilde{A} = i(A_3 - q)$, $\tilde{B} = i(p - A_2)$.

For the case (12.3) we get:

$$\Phi = \begin{pmatrix} A\Gamma \\ B\Gamma \end{pmatrix}, \tag{12.6}$$

where

$$A = \begin{pmatrix} 0 & \alpha \\ \mu & 0 \end{pmatrix}, \quad B = \begin{pmatrix} 0 & \beta \\ \mu\rho & 0 \end{pmatrix}, \quad \Gamma = \begin{pmatrix} \psi(t,x) \\ \phi(t,x) \end{pmatrix},$$

$$c = \frac{iq(\alpha^2+\beta^2)+2m\alpha\beta}{ip(\alpha^2+\beta^2)+m(\alpha^2-\beta^2)}, \quad \rho = \frac{i(\alpha p+\beta q)+\alpha m}{i(\alpha q-\beta p)+\beta m}. \tag{12.7}$$

The condition $c^* = c$ gives us for $\alpha \equiv \alpha_R + i\alpha_I$, $\beta \equiv \beta_R + i\beta_I$

$$\frac{\beta_I}{\alpha_I} = \frac{q\alpha_R - p\beta_R}{p\alpha_R + q\beta_R},$$

$$(p\alpha_R + q\beta_R)^2(\alpha_R{}^2 - \alpha_I{}^2 + \beta_R{}^2) - \alpha_I{}^2(q\alpha_R - p\beta_R)^2$$
$$+2m\alpha_I(\beta_R{}^2 + \alpha_R{}^2)(p\alpha_R + q\beta_R) = 0. \tag{12.8}$$

One of the nontrivial solutions of (12.8) is

$$\beta_I = 0, \quad \beta_R = \frac{q\alpha_R}{p}, \quad \alpha_I{}^{\pm} = \frac{\alpha_R}{p}(m \pm \sqrt{m^2+p^2+q^2}). \tag{12.9}$$

Substituting (12.6), (12.7) into (12.1) and taking (12.8) into account, we get a 2×2 equation (12.5) for Γ where $J = \sigma_3$,

$$U = \begin{pmatrix} 0 & u \\ v & 0 \end{pmatrix},$$

$$u(t,x) = \lambda_1 + \lambda_2 Q(t,x), \quad v(t,x) = \nu_1 + \nu_2 Q(t,x),$$

$$\lambda_1 = \frac{m\beta + i(q\alpha - p\beta)}{\mu}, \quad \lambda_2 = \frac{1}{\mu}\frac{(\alpha^2+\beta^2)(q\alpha - p\beta - im\beta)}{ip(\alpha^2+\beta^2)+m(\alpha^2-\beta^2)},$$
$$\nu_1 = \frac{\mu(m^2+q^2+p^2)}{i(p\beta - q\alpha) - m\beta}, \quad \nu_2 = \frac{i\mu}{\beta}(1 - c\rho). \tag{12.10}$$

Let χ_1 and χ_2 be 4×4 (for the case (12.2)) or 2×2 (for the case (12.3)) matrix solutions of equation (12.5). We define a matrix function $\tau_1 \equiv \chi_{1,x}\chi_1^{-1}$. It easy to see that τ_1 satisfies the following nonlinear equations:

$$\tau_{1,t} = \sigma_3\tau_{1,x} + [U, \tau_1] + [\sigma_3, \tau_1] + U_x.$$

Equation (12.5) is covariant with respect to DT:

$$\chi_2[1] = \chi_{2,x} - \tau_1\chi_2, \quad U[1] = U + [\sigma_3, \tau_1]. \tag{12.11}$$

It is necessary to choose the function χ_1 in such a way that the structure of the matrix $U[1]$ be the same as the structure of the matrix U. This is the condition that we call the reduction restriction.

The transformation (12.11) allows us to construct a superalgebra, in just the same way as the DT for the steady–state Schrödinger equation in the one-dimensional case [8]. In order to do this we introduce the following operators:

$$G^{(+)} = \frac{\partial}{\partial x} - \tau_1, \quad G^{(-)} = \frac{\partial}{\partial x} + \tau_1^+.$$

Let us define several new operators as follow:

$$h \equiv G^{(-)}G^{(+)}, \quad h[1] \equiv G^{(+)}G^{(-)},$$

$$T \equiv \frac{\partial}{\partial t} - J\frac{\partial}{\partial x} - U, \quad T[1] \equiv \frac{\partial}{\partial t} - J\frac{\partial}{\partial x} - U[1]. \tag{12.12}$$

It easy to see that

$$G^{(+)}T = T[1]G^{(+)}, \quad TG^{(-)} = G^{(-)}T[1], \quad [h,T] = [h[1],T[1]] = 0.$$

The operators h and $h[1]$ are the typical one–dimensional matrix Hamiltonians:

$$h = \frac{\partial^2}{\partial x^2} + \rho_D \frac{\partial}{\partial x} + V, \quad h[1] = \frac{\partial^2}{\partial x^2} + \rho_D \frac{\partial}{\partial x} + V[1],$$

$$\rho_D = (\tau_1^+ - \tau_1)_D, \quad V = -(\tau_{1,x} + \tau_1^+\tau_1), \quad V[1] = \tau_{1,x}^+ - \tau_1\tau_1^+,$$

where $(\tau_1)_D$ is the diagonal part of τ_1. It is easy to see that the operators $q^{(\pm)}$, H

$$q^{(+)} = \begin{pmatrix} 0 & 0 \\ G^{(+)} & 0 \end{pmatrix}, \quad q^{(-)} = \begin{pmatrix} 0 & G^{(-)} \\ 0 & 0 \end{pmatrix}, \quad H = \begin{pmatrix} h & 0 \\ 0 & h[1] \end{pmatrix}$$

generate a supersymmetry algebra:

$$\{q^{(\pm)}, q^{(\pm)}\} = [q^{(\pm)}, H] = 0, \quad \{q^{(+)}, q^{(-)}\} = H.$$

12.2 Crum Law

Let us consider $2N+1$ particular solutions of (12.5) with $\Gamma_k \equiv (\psi_k, \phi_k)^T$, $k \le 2N$, $\Phi \equiv (\psi, \phi)$:

$$\psi_{k,t} + \psi_{k,x} + u(t,x)\phi_k = 0, \quad \phi_{k,t} - \phi_{k,x} + v(t,x)\psi_k = 0. \tag{12.13}$$

The following theorem can be established:

Theorem 12.1. *The functions $\psi[N]$ and $\phi[N]$ defined below satisfy* (12.13) *with potentials $u[N]$ and $v[N]$ also defined below:*

$$\psi[N] = \frac{\Delta_1}{D}, \quad \phi[N] = \frac{\Delta_2}{D}, \quad u[N] = u + 2\frac{D_1}{D}, \quad v[N] = v - 2\frac{D_2}{D}, \tag{12.14}$$

where $\Delta_{1,2}$, $D_{1,2}$ *and* D *are the following determinants* ($\psi^{(N)} = \frac{\partial^N \psi(t,x)}{\partial x^N}$) :

$$D = \begin{vmatrix} \psi_1^{(N-1)} & \dots & \psi_1 & \phi_1^{(N-1)} & \dots & \phi_1 \\ \psi_2^{(N-1)} & \dots & \psi_2 & \phi_2^{(N-1)} & \dots & \phi_2 \\ \vdots & & \vdots & \vdots & & \vdots \\ \psi_{2N}^{(N-1)} & \dots & \psi_{2N} & \phi_{2N}^{(N-1)} & \dots & \phi_{2N} \end{vmatrix},$$

$$D_1 = \begin{vmatrix} \psi_1^{(N)} & \dots & \psi_1 & \phi_1^{(N-2)} & \dots & \phi_1 \\ \psi_2^{(N)} & \dots & \psi_2 & \phi_2^{(N-2)} & \dots & \phi_2 \\ \vdots & & \vdots & \vdots & & \vdots \\ \psi_{2N}^{(N)} & \dots & \psi_{2N} & \phi_{2N}^{(N-2)} & \dots & \phi_{2N} \end{vmatrix},$$

$$D_2 = \begin{vmatrix} \psi_1^{(N-2)} & \dots & \psi_1 & \phi_1^{(N)} & \dots & \phi_1 \\ \psi_2^{(N-2)} & \dots & \psi_2 & \phi_2^{(N)} & \dots & \phi_2 \\ \vdots & & \vdots & \vdots & & \vdots \\ \psi_{2N}^{(N-2)} & \dots & \psi_{2N} & \phi_{2N}^{(N)} & \dots & \phi_{2N} \end{vmatrix},$$

$$\Delta_1 = \begin{vmatrix} \psi^{(N)} & \dots & \psi & \phi^{(N-1)} & \dots & \phi \\ \psi_1^{(N)} & \dots & \psi_1 & \phi_1^{(N-1)} & \dots & \phi_1 \\ \vdots & & \vdots & \vdots & & \vdots \\ \psi_{2N}^{(N)} & \dots & \psi_{2N} & \phi_{2N}^{(N-1)} & \dots & \phi_{2N} \end{vmatrix},$$

$$\Delta_2 = \begin{vmatrix} \phi^{(N)} & \psi^{(N-1)} & \dots & \psi & \phi^{(N-1)} & \dots & \phi \\ \phi_1^{(N)} & \psi_1^{(N-1)} & \dots & \psi_1 & \phi_1^{(N-1)} & \dots & \phi_1 \\ \vdots & \vdots & & \vdots & \vdots & & \vdots \\ \phi_{2N}^{(N)} & \psi_{2N}^{(N-1)} & \dots & \psi_{2N} & \phi_{2N}^{(N-1)} & \dots & \phi_{2N} \end{vmatrix}.$$

Proof: We construct N, 2×2 matrix functions $\chi_k = (\Phi_{2k-1}, \Phi_{2k})$, $1 \le k \le N$:

$$\chi_k = \begin{pmatrix} \psi_{2k-1} & \psi_{2k} \\ \phi_{2k-1} & \phi_{2k} \end{pmatrix}.$$

These functions satisfy the matrix equation (12.5) with $J = \sigma_3$. After N–times DT (12.11), we get $\chi[N]$ and $U[N]$ which satisfy (12.5). Let us write $\chi[N]$ as a series:

$$\chi[N] = \chi^{(N)} - \sum_{i=1}^{N} A_i(t,x)\chi^{(N-i)}, \quad A_i = \begin{pmatrix} a_i & b_i \\ c_i & d_i \end{pmatrix}. \tag{12.15}$$

Substituting (12.15) into (12.13) we get (here $C_N^k = \frac{N!}{k!\,(N-k)!}$):

$$\sum_{k=0}^{N} C_N^k U^{(k)} \chi^{(N-k)} + \sum_{k=1}^{N} [(A_{k,t} - \sigma_3 A_{k,x})\chi^{(N-k)} - 2\sigma_3 (A_k)_F \chi^{(N-k+1)}]$$
$$+ \sum_{k=1}^{N} \sum_{i=0}^{N-k} C_N^i A_k U^{(i)} \chi^{(N-k-i)} - U[N](\chi^{(N)} + \sum_{k=1}^{N} A_k \chi^{(N-k)}) = 0,$$

where $(A_k)_F$ is the off–diagonal part of A_k. Therefore

$$u[N] = u + 2b_1, \quad v[N] = v - 2c_1.$$

To compute b_1 and c_1, we take into account that $\chi_k[N] = 0$ if $k \le N$, therefore we get a system of $2N$ equations as follows:

$$\psi_i^{(N)} = \sum_{n=1}^{N} (a_n \psi_i^{(N-n)} + b_n \phi_i^{(N-n)}), \quad \phi_i^{(N)} = \sum_{n=1}^{N} (c_n \psi_i^{(N-n)} + d_n \phi_i^{(N-n)}),$$

for $i = 1, \ldots, N$. Using Kramer's formulae, we get (12.14). □

In (12.10), if $\mu = i(\beta^* - c\alpha^*)^{-1}$, then ν_2 and λ_2 are real. The constants λ_1 and ν_1 may be annihilated by the standard gauge $U(1)$ transformations. Using the freedom in our choice of parameters, let us assume that $v(t,x) = \kappa u(t,x)$, $\kappa = \pm 1$. Now we can supplement (12.14) with the transformations law for χ_1 $(N = 1)$:

$$\psi_1[1] = \frac{\phi_2}{\Delta}, \quad \psi_2[1] = \frac{\kappa\phi_1}{\Delta}, \quad \phi_1[1] = \frac{\kappa\psi_2}{\Delta},$$
$$\phi_2[1] = \frac{\psi_1}{\Delta}, \quad \Delta = \begin{vmatrix} \psi_1 & \psi_2 \\ \phi_1 & \phi_2 \end{vmatrix}. \tag{12.16}$$

It is necessary to require that after N–times DT, the following reduction restriction will be true:

$$v[N] = \kappa u[N]. \tag{12.17}$$

In the general case, we do not have an algorithm allowing us to keep (12.17) in all the steps of DT. However, it is possible by the introduction of the so called *binary DT* which allows one to preserve the reduction restriction (12.17) [57].

Let us consider a closed 1–form

$$d\Omega = dx\,\zeta\chi + dt\,\zeta\sigma_3\chi, \quad \Omega \equiv \int d\Omega$$

where a 2×2 matrix function ζ solves the equation:

$$\zeta_t = \zeta_x \sigma_3 - \zeta U. \tag{12.18}$$

We shall apply the DT of (12.5). We can verify by immediate substitution that (12.18) is covariant with respect to the transform if

$$\zeta[+1] = \Omega(\zeta, \chi)\chi^{-1}.$$

Now we can transform U as

$$U[+1,-1] = U + [\sigma_3, \chi\Omega^{-1}\zeta]. \tag{12.19}$$

It can be shown that

$$\chi[+N,-N] = \chi - \sum_{k=1}^{N} \theta_k \Omega(\zeta_k, \chi), \quad \zeta[+N,-N] = \zeta - \sum_{k=1}^{N} \Omega(\zeta, \chi_k) s_k,$$

where θ_k and s_k can be found from the following equations:

$$\sum_{k=1}^{N} \theta_k \Omega(\zeta_k, \chi_i) = \chi_i, \quad \sum_{k=1}^{N} \Omega(\zeta_i, \chi_k) s_k = \zeta_i.$$

The transformation:

$$U[+N,-N] = U + \sum_{i,k=1}^{N} [\sigma_3, \theta_i \Omega(\zeta_k, \chi_i) s_k]$$

is the forementioned *binary* DT.

Let $v = \kappa u^*$, then $U[+N,-N]$ will satisfy the reduction restriction if we choose ζ_k and χ_k such that:

$$\zeta_k = \chi_k R, \quad R = \mathrm{diag}(1, -\kappa).$$

The solution that follows from (12.19) has the form

$$u[+1,-1] = u + \frac{2\kappa(\psi_2\phi_1^*\theta_{12}^* + \psi_1\phi_2^*\theta_{12} - \psi_1\phi_1^*\theta_{22} - \psi_2\phi_2^*\theta_{11})}{\theta_{11}\theta_{22} - |\theta_{12}|^2},$$

$$\theta_{ik} = \int dx\,(\psi_i^*\psi_k - \kappa\phi_i^*\phi_k) + dt\,(\psi_i^*\psi_k + \kappa\phi_i^*\phi_k).$$

Note that the square of the absolute value of $u[+1,-1]$ can be expressed by the concise formula:

$$|u[+1,-1]|^2 = |u|^2 - \kappa\left(\frac{\partial^2}{\partial t^2} - \frac{\partial^2}{\partial x^2}\right)\ln(\theta_{11}\theta_{22} - |\theta_{12}|^2).$$

DT allows us to construct a rich set of exact solutions of the Dirac equation with (1+1) potentials. Finally, we consider fields decreasing as $t, x \to \pm\infty$ (adiabatic engaging and turning-off). Let $u = v = 0$. The particular solutions of (12.13) are:

$$\psi_k = A_k\, e^{\omega(t-x)} + B_k\, e^{\omega(x-t)}, \quad \phi_k = C_k\, e^{\lambda(t+x)} + D_k\, e^{-\lambda(t+x)},$$

where $k = 1, 2$; A, B, C, D, ω and λ are real constants. After DT (12.11) under condition (12.10), we get:

$$Q(t, x) = \frac{1}{\mu_1 \cosh(at + bx + \delta_1) + \mu_2 \cosh(bt + ax + \delta_2)}, \tag{12.20}$$

$$u(t, x) = \xi\, Q(t, x), \quad v(t, x) = -\frac{1 + c^2}{\xi^2} u(t, x),$$

$$\xi = |\alpha c - \beta|^2 = 4\omega(A_1 B_2 - A_2 B_1), \quad a = \omega + \lambda, \quad b = \lambda - \omega, \quad \mu(\beta^* - c\alpha^*) = i,$$

$$1 + c^2 = 4\lambda\xi(C_2 D_1 - C_1 D_2), \quad \omega(A_1 B_2 - A_2 B_1) = \lambda(D_1 C_2 - D_2 C_1),$$

$$\mu_1 = \sqrt{(A_1 C_2 - A_2 C_1)(B_1 D_2 - B_2 D_1)}, \quad \mu_2 = \sqrt{(B_1 C_2 - B_2 C_1)(A_1 D_2 - A_2 D_1)},$$

$$\delta_1 = \frac{1}{2} \ln \frac{A_1 C_2 - A_2 C_1}{B_1 D_2 - B_2 D_1}, \quad \delta_2 = \frac{1}{2} \ln \frac{B_1 C_2 - B_2 C_1}{A_1 D_2 - A_2 D_1},$$

where α and β satisfy (12.9). The potential (12.20) is a localized impulse which decreases as $t \to \pm\infty$. A particular solution of the Dirac equation may be calculated from $(\psi_{1,2}[1], \phi_{1,2}[1])^T$, using (12.4) and (12.6), where $\psi_{1,2}[1]$ and $\phi_{1,2}[1]$ are defined in (12.16). This bispinor describes the quasi-stationary state of a fermion. The fermion is free along (y, z)-axes and restrained along x-axis.

Chapter 13

Moutard Transformations for the 2D and 3D Schrödinger Equations

Darboux transformations (DT) for 1D Schrödinger equation is well-known. In this chapter, we will investigate DT for 2D and 3D Schrödinger equations. For instance, the Novikov-Veselov equation (NVE) can be represented as a compatibility condition of the linear system

$$\hat{H}\psi = -\psi_{xx} - \psi_{yy} + U\psi = 0, \quad \psi_t = A\psi, \tag{13.1}$$

where $\psi = \psi(x, y, t)$, $U = U(x, y, t)$, and A is a linear operator. The system (13.1) is covariant with respect to the action of Darboux type transformation which should be more properly called a Moutard transformation (MT) since both the prop and the dressed eigenfunctions are at the same spectral value (here it is zero). The MT may not be as powerful as DT in the quantum theory (i.e., we can not use it to find spectrum), but it is very useful for constructing exact solutions of the NVE.

13.1 A 2D Moutard Transformation

Let's consider the 2D Schrödinger (13.1) $\hat{H}\psi = 0$ with $U = U(x, y)$, $\psi = \psi(x, y)$. Let $\varphi = \varphi(x, y)$ be another solution: $\hat{H}\varphi = 0$ such that $\psi \neq \varphi$. We define operators $\hat{q}_i = \partial_i - \partial_i \ln\varphi$ and $\hat{q}_i^+ = -\partial_i - \partial_i \ln\varphi$, with $i = 1, 2$. Then

$$\hat{H} = \hat{q}_m^+ \hat{q}_m = -\Delta_{(2)} + U,$$

and summation is implied over repeated indices. Let's define the "dressed" Hamiltonian $\hat{\tilde{H}}$

$$\hat{\tilde{H}} = \hat{q}_m \hat{q}_m^+ = -\Delta_{(2)} + \tilde{U},$$

where

$$\tilde{U} = U - 2\Delta_{(2)} \ln\varphi. \tag{13.2}$$

Then one needs to find such $\tilde{\psi}$ that

$$\hat{\tilde{H}}\tilde{\psi} = -\Delta_{(2)}\tilde{\psi} + \tilde{U}\tilde{\psi} = 0. \tag{13.3}$$

A solution is known:

$$\tilde{\psi} = \frac{\theta}{\varphi}, \tag{13.4}$$

where $\theta = \int d\theta$,

$$d\theta = dx^m \varepsilon_{mk}\varphi\partial_k\psi = dx^1\Omega_1 + dx^2\Omega_2. \tag{13.5}$$

The map $U \to \tilde{U}$, $\psi \to \tilde{\psi}$ is the MT. Another way to see this MT is as follows. Notice that $\tilde{\psi}$ satisfies

$$\hat{\tilde{H}}\tilde{\psi} = \hat{q}_m\hat{q}_m^+\tilde{\psi} = 0, \tag{13.6}$$

Notice also that

$$\varepsilon_{mk}\hat{q}_m\hat{q}_k = 0 \Longrightarrow \varepsilon_{mk}\hat{q}_m\hat{q}_k\psi = 0. \tag{13.7}$$

Thus

$$\hat{q}_m\hat{q}_m^+\tilde{\psi} = \varepsilon_{mk}\hat{q}_m\hat{q}_k\psi \Longrightarrow \hat{q}_m(\hat{q}_m^+\tilde{\psi} - \varepsilon_{mk}\hat{q}_k\psi) = 0.$$

One can choose

$$\hat{q}_m^+\tilde{\psi} = \varepsilon_{mk}\hat{q}_k\psi.$$

This equation can be write in another way:

$$\partial_m(\tilde{\psi}\varphi) = \varepsilon_{km}(\varphi\partial_k\psi - \psi\partial_k\varphi). \tag{13.8}$$

Equation (13.8) is true only if the curl of the right is zero. Then one gets

$$(\varepsilon_{km}\partial_l - \varepsilon_{kl}\partial_m)(\varphi\hat{q}_k\psi) = 0. \tag{13.9}$$

It's easy to show that the last equation is true if and only if, ψ and φ are solutions of the equations $\hat{H}\psi = 0$ and $\hat{H}\varphi = 0$ correspondingly. If we integrate equation (13.8), then we get the MT (13.4) and (13.5).

13.2 A 3D Moutard Transformation

The Hamiltonian in 3D has the form:

$$\hat{H} = -\Delta_{(3)} + U(x, y, z) = -\partial_1^2 - \partial_2^2 - \partial_3^2 + U(x_1, x_2, x_3). \tag{13.10}$$

As before, one can write

$$\hat{H}\psi = 0 \tag{13.11}$$

with

$$\hat{H} = \hat{q}_m^+\hat{q}_m, \quad \hat{q}_m \equiv \partial_m - \partial_m \ln\varphi,$$

where $\hat{H}\varphi = 0$, $m = 1, 2, 3$. The dressed Hamiltonian will be

$$\hat{\tilde{H}} \equiv \hat{q}_m \hat{q}_m^+ = -\Delta_{(3)} + \tilde{U} = -\Delta_{(3)} + U - 2\Delta_{(3)} \ln \varphi.$$

Now one needs to find such $\tilde{\psi}$ that

$$\hat{\tilde{H}} \tilde{\psi} = 0.$$

Using the simple method above for 2D, one gets instead of (13.8)

$$\hat{q}_m^+ \tilde{\psi} = \varepsilon_{mkl} \hat{q}_k \hat{q}_l^+ \psi, \tag{13.12}$$

and instead of (13.9)

$$(\varepsilon_{lkm} \partial_s - \varepsilon_{skm} \partial_l)(\varphi \hat{q}_k \hat{q}_m^+ \psi) = 0 \tag{13.13}$$

Unfortunately, in contrast to (13.9), equation (13.13) is not true if ψ and φ are solution of (13.11). This is the principal problem in getting MT in 3D (and $D > 3$ as well). One can obtain MT in 3D by introducing an external magnetic field with magnetic induction $\vec{B} = \text{rot}\ \vec{\kappa}$, where $\vec{\kappa}$ is the vectorial potential. Of course, this field is a solution of the Maxwell equation

$$\text{div} \vec{B} = 0, \tag{13.14}$$

The introduction of the magnetic field can be described by the usual rule

$$\partial_m \to D_m = \partial_m + i\kappa_m.$$

Define the superpotential $\chi = \chi(x, y, z) \equiv -\ln \varphi$. Then one gets the following Hamiltonians:

$$\hat{H} = \hat{Q}_m^+ \hat{Q}_m, \qquad \hat{\tilde{H}} = \hat{Q}_m \hat{Q}_m^+,$$

where:

$$\begin{aligned} \hat{Q}_m &= \partial_m + \theta_m, \quad \hat{Q}_m^+ = -\partial_m + \theta_m^*, \\ \theta_m &= i\kappa_m + \partial_m \chi, \quad \theta_m^* = -i\kappa_m + \partial_m \chi, \end{aligned} \tag{13.15}$$

It is easy to show that

$$\varepsilon_{mkl} \hat{q}_m \hat{q}_k \hat{q}_l^+ = i(\vec{B}, -\vec{\nabla} - i\vec{\kappa} + \text{grad}\ \chi).$$

Thus, it is identically zero if

$$\vec{B} = \text{rot}\, \vec{\kappa} = \vec{0}. \tag{13.16}$$

Below we will consider the Aaronov-Bohm magnetic field: $\vec{B} = 0$, $\vec{\kappa} \neq 0$. We have

$$\hat{H} = -\Delta_{(3)} - i\text{div}\, \vec{\kappa} - 2i(\vec{\kappa}, \vec{\nabla}) + \vec{\kappa}^2 + \frac{\Delta \varphi}{\varphi}, \tag{13.17}$$

and

$$\hat{\tilde{H}} = -\Delta - i\mathrm{div}\,\vec{\kappa} - 2i(\vec{\kappa}, \vec{\nabla}) + \vec{\kappa}^2 - \frac{\Delta\varphi}{\varphi} + 2\left(\frac{\vec{\nabla}\varphi}{\varphi}\right)^2. \tag{13.18}$$

Notice also that $\hat{H} = -\Delta_{(3)} + U$ and $\hat{\tilde{H}} = -\Delta_{(3)} + \tilde{U}$. We have

$$\tilde{U} = U - 2\Delta_{(3)} \ln \varphi. \tag{13.19}$$

When $\vec{B} = \vec{0}$, we get from (13.12)

$$\hat{Q}_m^+ \tilde{\psi} = \varepsilon_{mkl} \hat{Q}_k \hat{Q}_l^+ \psi,$$

By analogy with the 2D case, we shall find $\tilde{\psi}$ as

$$\tilde{\psi} = \frac{f}{\varphi} = e^{\chi} f, \tag{13.20}$$

where f satisfies

$$\vec{\nabla} f + i\vec{\kappa} f = 2\vec{\nabla}\psi \wedge \vec{\nabla}\varphi + 2i\vec{\kappa} \wedge \vec{\nabla}\varphi\psi. \tag{13.21}$$

Define p as

$$\vec{\kappa} = -i\vec{\nabla} p, \tag{13.22}$$

then (13.21) has the form

$$\vec{\nabla} f + f\vec{\nabla} p = 2\left(\vec{\nabla}\psi + \psi\vec{\nabla} p\right) \wedge \vec{\nabla}\varphi. \tag{13.23}$$

Solving this equation, one gets

$$\tilde{\psi} = \frac{f}{\varphi} = \frac{e^{-p}}{\varphi}\left(2\int_{\vec{r}_0}^{\vec{r}} \left(\vec{\nabla}(e^p\psi), \vec{\nabla}\varphi, d\vec{r}\right) + A\right), \tag{13.24}$$

where A is an arbitrary constant. The integrand has to be a gradient, therefore, we get the additional requirement on p (and on κ as well):

$$\vec{\nabla} \wedge \left(\vec{\nabla}(e^p\psi) \wedge \vec{\nabla}\varphi\right) = \vec{0}. \tag{13.25}$$

Direct substitution shows that such $\tilde{\psi}$ is a solution to the equation $\hat{\tilde{H}}\tilde{\psi} = 0$, with $\hat{\tilde{H}}$ given by (13.18). The transformations (13.19) and (13.24) are the 3D Moutard transformations.

Chapter 14

BLP Equation

Like Davey-Stewartson (DS) and Kadomtsev-Petviashvili (KP) equations, the Boiti-Leon-Pempinelli (BLP) equation is another (1+2)-dimensional integrable system [20, 37]. A somewhat peculiar feature of the BLP equation is that a reduction is the viscous Burgers equation. Since BLP equation has to be a Hamiltonian system and viscous Burgers equation is a dissipative system, it seems to be paradoxical. The answer to the puzzle is that the Hamiltonian of the BLP equation reduces to 0 by this reduction.

14.1 The Darboux Transformations for the BLP Equation

The BLP equation can be written in the form

$$\begin{aligned} &a_{ty} + (a^2 - a_x)_{xy} + 2b_{xx} = 0, \\ &b_t + (b_x + 2ab)_x = 0, \end{aligned} \tag{14.1}$$

where $a = a(t, x, y)$ and $b = b(t, x, y)$ are real-valued functions of three variables. One can rewrite (14.1) in the following form with the change of variable $b = p_y$,

$$\begin{aligned} &a_t + 2aa_x + (2p - a)_{xx} = 0, \\ &p_{yt} + (p_{yx} + 2ap_y)_x = 0. \end{aligned} \tag{14.2}$$

The reduction $p = 0$ reduces the BLP equation (14.2) to the viscous Burgers equation

$$a_t + 2aa_x - a_{xx} = 0. \tag{14.3}$$

Another reduction $p = a$ reduces the BLP equation (14.2) to the "anti-Burgers equation".

The Lax pair of the BLP equation is

$$\psi_{xy} + a\psi_y + p_y\psi = 0, \quad \psi_t = \psi_{xx} + 2p_x\psi. \tag{14.4}$$

Theorem 14.1. *Let ϕ and ψ be two solutions of (14.4). Define two functions $\tau = \partial_x(\ln\phi)$ and $\rho = (\partial_y(\ln\phi))^{-1}$. The Lax pair (14.4) is covariant with respect to the two types of Darboux transformations,*

$$\psi \to \psi[1] = \rho\psi_y - \psi, \quad a \to a[1] = a - \partial_x \ln \rho, \quad p \to p[1] = p - \rho p_y, \tag{14.5}$$

and

$$\psi \to \psi[1] = \psi_x - \tau\psi, \quad a \to a[1] = a - \partial_x \ln(a+\tau), \quad p \to p[1] = p + \tau. \tag{14.6}$$

In addition, it is also covariant with respect to the Laplace transformation,

$$\psi \to \psi[1] = \frac{\psi_y}{p_y}, \ a \to a[1] = a + \partial_x \ln p_y, \ p \to p[1] = p + a + \partial_x \ln p_y, \tag{14.7}$$

and the Laplace inverse transformation is given by

$$\psi \to \psi[-1] = \partial_x\psi + a\psi, \ a \to a[-1] = a - \partial_x \ln(p-a)_y, \ p \to p[-1] = p - a. \tag{14.8}$$

Let us define three functions u, v, χ by,

$$a = -\partial_x \ln u, \quad b = p_y = -uv, \quad \chi = \frac{\partial_x \psi}{u}$$

and cone variables ξ and η,

$$\partial_y = \partial_\eta - \partial_\xi, \quad \partial_x = \partial_\eta + \partial_\xi,$$

then the spatial part of the Lax pair (14.4) can be rewritten in the Zakharov-Shabat form

$$\partial_\eta \Phi = \sigma_3 \partial_\xi \Phi + U\Phi, \tag{14.9}$$

where

$$\Phi = \begin{pmatrix} \psi \\ \chi \end{pmatrix}, \quad U = \begin{pmatrix} 0 & u \\ v & 0 \end{pmatrix}, \quad \sigma_3 = \begin{pmatrix} 1 & 0 \\ 0 & -1 \end{pmatrix}.$$

Equation (14.9) is the spectral problem for the DS equations. The above Laplace transformation then produces an explicitly invertible Bäcklund autotransformation for the DS equations, which can be used to construct solutions to the DS equations that fall off in all directions in the plane according to exponential and algebraic law.

14.2 Crum Law

Consider the DT (14.5) and define $Q_N\psi = \partial_y^{N+1}\partial_x\psi$. It is easy to check that if ψ is a solution of (14.4), then Q_N can be written as

$$Q_N = -a\partial_y^{N+1} - (p+Na)_y\partial_y^N + \sum_{k=1}^{N-1} c_{N,k}\partial_y^k - \partial_y^{N+1}p, \tag{14.10}$$

where $c_{N,k} = c_{N,k}(a, p_y; a_x, p_{xy}; a_y, p_{yy}; \ldots)$. Let $\{\psi, \psi_1, \ldots, \psi_N\}$ be $N+1$ particular solutions of (14.4). After N times of DT (14.5), we get

$$\psi[N] = -\psi + \sum_{k=1}^{N} \alpha_{N,k}\partial_y^k \psi. \tag{14.11}$$

Substituting (14.11) into

$$\psi_{xy}[N] + a[N]\psi_y[N] + p_y[N]\psi[N] = 0 \tag{14.12}$$

and using (14.10), we get

$$a[N] = a - \partial_x \ln \alpha_{N,N}, \quad p[N] = p - \sum_{k=1}^{N} \alpha_{N,k}\partial_y^k p.$$

To compute $\alpha_{N,k}$, we take into account that

$$\sum_{k=1}^{N} \alpha_{N,k}\partial_y^k \psi_j = \psi_j, \quad j = 1, \ldots, N.$$

Thus

$$\begin{aligned} \psi[N] = \frac{D_N(1|1)}{D_N(1, N+2|1,2)}, \quad a[N] = a - \partial_x \ln \frac{D_N(1,2|1,2)}{D_N(1, N+2|1,2)}, \\ p[N] = p + (-1)^N \frac{D_N(1|2)}{D_N(1, N+2|1,2)}, \end{aligned} \tag{14.13}$$

where $D_N(i, j|k, m)$ are determinants which can be obtained by the deletion of columns i, j and rows k, m from the $(N+2) \times (N+2)$ determinant D_N

$$D_N = \begin{vmatrix} \partial^{N+1} p & \partial^N p & \partial^{N-1} p & \ldots & \partial p & 0 \\ \partial^{N+1}\psi & \partial^N \psi & \partial^{N-1}\psi & \ldots & \partial\psi & \psi \\ \partial^{N+1}\psi_1 & \partial^N \psi_1 & \partial^{N-1}\psi_1 & \ldots & \partial\psi_1 & \psi_1 \\ \vdots & \vdots & \vdots & & \vdots & \vdots \\ \partial^{N+1}\psi_N & \partial^N \psi_N & \partial^{N-1}\psi_N & \ldots & \partial\psi_N & \psi_N \end{vmatrix}, \tag{14.14}$$

where $\partial = \partial_y$.

Next we consider the Crum law for the DT (14.6). Define $P_N\psi = \partial_x^{N+1}\partial_y\psi$, then

$$P_N = -p_y\partial_x^N + \sum_{k=0}^{N-1} \beta_{N,k}\partial_x^k + \lambda_N\partial_y. \tag{14.15}$$

Since $P_{N+1} = \partial_x P_N$,

$$\begin{aligned} &\beta_{N+1,N} = \beta_{N,N-1} - p_{xy}, \quad \beta_{N+1,k} = \partial_x\beta_{N,k} + \beta_{N,k-1}, \quad 1 \le k \le N-1, \\ &\beta_{N+1,0} = \partial_x\beta_{N,0} - p_y\lambda_N, \quad \lambda_{N+1} = \partial_x\lambda_N - a\lambda_N. \end{aligned} \tag{14.16}$$

After N times of DT (14.6), we get

$$\psi[N] = \partial_x^N \psi + \sum_{k=0}^{N-1} \alpha_{N,N-k} \partial_x^k \psi. \tag{14.17}$$

Substituting (14.17) into (14.12) and taking (14.15) and (14.16) into account, we get

$$\begin{gathered} a[N]p_y[N] = ap_y + \partial_y \left(\partial_x(\alpha_{N,1} - Np) + \alpha_{N,2} - \frac{1}{2}\alpha_{N,1}^2 \right), \\ p[N] = p - \alpha_{N,1}. \end{gathered} \tag{14.18}$$

To compute $\alpha_{N,k}$, we notice that $(j = 1, \ldots, N)$

$$\partial_x^N \psi_j + \sum_{k=0}^{N-1} \alpha_{N,N-k} \partial_x^k \psi_j = 0, \tag{14.19}$$

therefore

$$\psi[N] = \frac{W_N}{W_N(1|1)}, \qquad \alpha_{N,1} = -\frac{W_N(2|1)}{W_N(1|1)}, \qquad \alpha_{N,2} = \frac{W_N(3|1)}{W_N(1|1)}, \tag{14.20}$$

where

$$W_N = \begin{vmatrix} \partial^N \psi & \partial^{N-1}\psi & \ldots & \psi \\ \partial^N \psi_1 & \partial^{N-1}\psi_1 & \ldots & \psi_1 \\ \vdots & \vdots & & \vdots \\ \partial^N \psi_N & \partial^{N-1}\psi_N & \ldots & \psi_N \end{vmatrix} \tag{14.21}$$

and $\partial = \partial_x$.

Starting from (14.7) and (14.8), we can also find the N-fold LT $a[N]$ $(a[-N])$ and $b[N]$ $(b[-N])$. (it is more convenient to use field b instead of field p in this case.) After N times of LT (14.7), we get

$$\psi[N] = \sum_{k=1}^{N} \alpha_{N,k} \partial_y^k \psi. \tag{14.22}$$

For simplicity, we denote $a[N]$ by a_N and $b[N]$ by b_N. Substituting (14.22) into (14.12), we get

$$a_N = a - \partial_x \ln \alpha_{N,N}, \qquad b_N = b + \partial_y \left(Na - \partial_x \ln(\alpha_{N-1,N-1}\alpha_{N,N}) \right) \tag{14.23}$$

with the supplementary condition

$$\sum_{k=1}^{N} \alpha_{N,k} \partial_y^{k-1} b = 0. \tag{14.24}$$

Thus

$$\alpha_{N,N} = \left(\prod_{j=1}^{N} b_{j-1} \right)^{-1}, \tag{14.25}$$

and (14.23) is the nonlinear superposition formula for the BLP equations. To compute $\psi[N]$ we must obtain the rest coefficients $\alpha_{N,k}$. We need to introduce the operators

$$\mathbf{f}_j = \frac{1}{b_j}\partial_y, \quad \mathbf{F} = \mathbf{f}_{N-1}\mathbf{f}_{N-2}\cdots\mathbf{f},$$

where $j = 0, \ldots, N-1$, $b_0 = b$, $\mathbf{f}_0 = \mathbf{f}$. It's clear that

$$\psi[N] = \sum_{k=1}^{N} \alpha_{N,k}\partial_y^k\psi - \mathbf{F}\psi. \tag{14.26}$$

The coefficients $\alpha_{N,k}$ from (14.26) can be calculated by finding N functions θ_j ($j = 1, \ldots, N$) such that $\mathbf{F}\theta_j = 0$. It is easy to obtain these functions in the following way: For an arbitrary index j we construct the sequence of equations for the j functions $\theta_j^{(k)}$ ($k = 1, \ldots, j$),

$$\mathbf{f}_{k-1}\theta_j^{(k-1)} = \theta_j^{(k)}, \quad \theta_j^{(0)} = \theta_j, \quad \theta_j^{(j-1)} = 1, \tag{14.27}$$

The system (14.27) can be solved and we get N functions θ_j,

$$\theta_j = < b < b_1 < \cdots < b_{j-1} > \cdots >_y,$$

where and in the following $< \cdots >_{x,y}$ denotes the integration in the x, y-variables

$$< S >_x = \frac{1}{2}\int dz \ \mathrm{sgn}(x-z)S(z,y,t), \quad < S >_y = \frac{1}{2}\int dz \ \mathrm{sgn}(y-z)S(x,z,t).$$

Using (14.24) to express $\alpha_{N,1}$, then

$$\mathbf{F}\theta_j = \sum_{k=2}^{N} M_{kj}\alpha_{N,k} = 0, \quad M_{kj} = \partial_y^k\theta_j - \frac{\partial_y^{k-1}b}{b}\partial_y\theta_j. \tag{14.28}$$

It is obvious that $M_{k1} = M_{k2} = 0$. Substituting (14.25) into (14.28), we get the nonhomogeneous system of linear equations for the desired coefficients, which can be solved by the Cramer's formulae.

The N-fold LT (14.8) can be obtained in analogy with the LT (14.7). It is easy to see that $\psi[-N]$ has the form (14.17) with new coefficients $\alpha_{N,k}$. This means that $b[-N]$ and $a[-N]$ can be computed by (14.18). Then we need to find the new coefficients $\alpha_{N,k}$. These coefficients may be expressed in terms of the determinants (14.20) and (14.21) with the change $\psi_k \to \theta_k$ where the functions (θ_k) satisfy equation (14.19). To find θ_k, we rewrite (14.19) as

$$\left(\partial_x + a_{N-1}\right)\left(\partial_x + a_{N-2}\right)\cdots\left(\partial_x + a\right)\theta_j = 0, \quad j = 1, \ldots, N \tag{14.29}$$

where we denote $a[-k]$ by a_k and $b[-k]$ by b_k. Then (14.29) can be rewritten as,

$$(\partial_x + a_j)\,\theta_m^{(j)} = \theta_m^{(j+1)}, \quad j = 0, \ldots, m-2,$$
$$(\partial_x + a_{m-1})\,\theta_m^{(m-1)} = 0, \tag{14.30}$$

for any θ_m. In (14.30), $\theta_m^{(0)} = \theta_m$, $m = 2, \ldots, N$, and

$$\theta_1 = \exp\left(- < a >_x\right).$$

By solving (14.30), we get the required functions

$$\theta_m = \theta_1\,\Phi_m, \quad \Phi_m =< b_1 < b_2 < \cdots < b_{m-1} > \cdots >_x .$$

14.3 Exact Solutions

In [37], Bäcklund transformations were applied to construct singular solutions that decay according to rational laws in all directions on the plane with $a \to 0$ and $b \to$ const. as $x^2 + y^2 \to \infty$. Here we shall consider two examples: (i) the localized nonsingular solution b decaying according to the exponential law as a function of x and according to the rational law as a function of y; (ii) the "blow-up" solution which is nonsingular when $t > 0$ and singular when $t < 0$.

Let $a = b = 0$. To construct the solution (i), we choose the solution of (14.4) in the form

$$\phi = \exp\left(\mu^2 t\right)\cosh\left(\mu x\right) + B(y),$$

where $B = B(y)$ is an arbitrary differentiable function. Taking (14.6) into account, we get

$$b[1] = -\frac{\mu B' \sinh\left(\mu x\right)\exp(\mu^2 t)}{\left(B + \cosh\left(\mu x\right)\exp(\mu^2 t)\right)^2},$$
$$a[1] = -\frac{\mu(\exp(\mu^2 t) + B\cosh\left(\mu x\right))}{\sinh\left(\mu x\right)\left(B + \cosh\left(\mu x\right)\exp\left(\mu^2 t\right)\right)}. \tag{14.31}$$

If we choose the function $B(y)$ to be increasing according to the exponential law as $y \to \pm\infty$, then the solution (14.31) has nonclosed level curves. To avoid this, we choose $B(y)$ to be an even polynomial (the property of being even guarantees the nonsingular behavior of (14.31)). Let $B(y) = \nu^2 y^{2N} + \lambda^2$ (ν and λ are constants and $\Im\nu = \Im\lambda = 0$), we get the desired solution

$$b[1] = -\frac{2\mu N \nu^2 y^{2N-1}\sinh\left(\mu x\right)\exp(\mu^2 t)}{(\lambda^2 + \nu^2 y^{2N} + \cosh\left(\mu x\right)\exp(\mu^2 t))^2}. \tag{14.32}$$

The regular solution (14.32) decays according to the exponential law as $x \to \pm\infty$, and the rational law $(1/y^{2N+1})$ as $y \to \pm\infty$. On the other hand, the singular solution $a[1]$ has level curves along y-lines.

To construct blow-up solutions of BLP, we shall apply two DTs, (14.5) and (14.6), on the vacuum background ($a = b = 0$). As a result, we get

$$b[2] = \frac{W_2(1|1)\, D_2(1,4|1,2)}{D_2^2(1,1|1,1)}. \tag{14.33}$$

The functions ψ_k ($k = 1,2$) in the determinants D and W (see (14.14) and (14.21)) are solutions of (14.4). We choose them to be

$$\psi_1 = C_1 \exp\left(\lambda^2 t\right)\cosh(\lambda x) + \cosh(\alpha y), \quad \psi_2 = C_2 \exp\left(-\mu^2 t\right)\sin(\mu x) + \sinh(\beta y).$$

Then

$$D_2(1,1|1,1) = D_1(y) + D_2(x,y,t)$$

where

$$D_1(y) = \beta \cosh(\beta y)\cosh(\alpha y) - \alpha \sinh(\beta y)\sinh(\alpha y),$$
$$D_2(x,y,t) = \beta C_1 \exp\left(\lambda^2 t\right)\cosh(\lambda x)\cosh(\beta y) - \alpha C_2 \exp\left(-\mu^2 t\right)\sinh(\alpha y)\sin(\mu x).$$

If $\beta > \alpha > 0$, then $D_1(y) > 0$ for $y \in (-\infty, +\infty)$. If $C_1 > 0$, then $D_2(x,0,t) > 0$. So the solution (14.33) is nonsingular at $y = 0$. A singularity can appear if $D_2(x,y,t) < 0$. It is possible that $D_2(x',y',t') = 0$, i.e.,

$$\rho \exp((\lambda^2 + \mu^2)t')\cosh(\beta y')\cosh(\lambda x') = \sinh(\alpha y')\sin(\mu x'), \tag{14.34}$$

when $\rho \equiv \beta C_1(\alpha C_2)^{-1}$. We can rewrite (14.34) as,

$$\rho \exp((\lambda^2 + \mu^2)t')|\theta_1(y')| = |\theta_2(x')|, \tag{14.35}$$

with

$$\theta_1(y) = \frac{\cosh(\beta y)}{\sinh(\alpha y)}, \quad \theta_2(x) = \frac{\sin(\mu x)}{\cosh(\lambda x)}.$$

Let y_0 and x_0 be solutions of the equations

$$\tanh(\beta y_0)\tanh(\alpha y_0) = \frac{\alpha}{\beta}, \quad \tanh(\lambda x_0)\tan(\mu x_0) = \frac{\mu}{\lambda}.$$

Since $\exp((\lambda^2 + \mu^2)t') \geq 1$ when $t' \geq 0$, for such values of t we can to choose

$$\rho > \frac{\theta_2(x_0)}{\theta_1(y_0)}. \tag{14.36}$$

Then the condition (14.36) guarantees the nonsingular behavior of (14.33) when $t \geq 0$. It is possible to choose parameters such that (14.33) is singular when $t < 0$. We choose ψ_2 as

$$\psi_2 = C_2 \exp(\mu^2 t)\sinh(\mu x) + \sinh(\beta y).$$

The solution (14.33) will be regular when $t \geq 0$, and singular when $t < 0$, if $\lambda > \mu$ and

$$\rho > \frac{\tilde{\theta}_2(\tilde{x}_0)}{\theta_1(y_0)}, \quad \tilde{\theta}_2(x) = \frac{\sinh(\mu x)}{\cosh(\lambda x)},$$

where $\tilde{x}_0 > 0$ is maximum point of the function $\tilde{\theta}_2(x)$.

14.4 Dressing From Burgers Equation

Starting from an arbitrary solution of the Burgers equation, one can construct exact solutions of (14.2) via LT (14.8). It is convenient to redefine variables in the following way

$$x = \frac{\xi}{\sqrt{\nu}}, \quad y = \frac{\eta}{\nu\sqrt{\nu}}, \quad a = \frac{A(\xi,\eta,t)}{2\sqrt{\nu}}, \quad p = \frac{P(\xi,\eta,t)}{\nu\sqrt{\nu}},$$

where $\nu > 0$ and μ are arbitrary constants. As a result, the BLP equations take the form

$$\begin{aligned} A_t + AA_\xi - \nu A_{\xi\xi} &= -4P_{\xi\xi}, \\ P_{t\eta} + (AP_\eta + \nu P_{\xi\eta})_\xi &= 0, \end{aligned} \tag{14.37}$$

where ν is the parameter that may be called the coefficient of viscosity. Assume that $P = 0$ and A is a solution of the Burgers equation, then (14.8) allows us to construct solutions of the BLP equations (14.37),

$$P[-1] = -\frac{\nu}{2}A, \quad A[-1] = A - 2\nu\partial_\xi \ln(A_\eta)\,. \tag{14.38}$$

Let us consider a simple example: The shock wave solution of the Burgers equation [105],

$$A = A_1 + \frac{A_2 - A_1}{1 + \exp(\frac{A_2 - A_1}{2\nu}(\xi - Ut))},$$

where

$$A_1 = A_1(\eta), \quad A_2 = A_2(\eta), \quad U = \frac{A_1 + A_2}{2}.$$

Assuming that $A_1 = 0$ and taking (14.38) into account, we get

$$\begin{aligned} A[-1] &= \frac{A_2\left[G^2 A_2(\xi - A_2 t) - 2\nu(2G^2 + 3G + 1)\right]}{(1+G)\left[GA_2(\xi - A_2 t) - 2\nu(1+G)\right]}, \\ P[-1] &= -\frac{\nu A_2}{2(1+G)}, \quad G = \exp\left(\frac{A_2(2\xi - A_2 t)}{4\nu}\right). \end{aligned} \tag{14.39}$$

It is interesting to consider the behavior of (14.39) as $\nu \to 0$. Assume that $A_2 > 0$ for all values of η. Then for $2\xi \neq A_2 t$, we get

$$P[-1] \to 0, \quad A[-1] \to A_2,$$

and for $2\xi = A_2 t$,

$$P[-1] \to 0, \quad A[-1] \to \frac{A_2}{2}.$$

Using other well known solutions of the Burgers equation (see [105]), it is easy to construct a rich set of exact solutions for the BLP equations via the above dressing procedure. For example, we consider the solution of the Burgers equation,

$$A = -2\nu\partial_\xi \ln f, \quad f = \alpha + \sqrt{\frac{\beta}{t}}\exp\left(-\frac{\xi^2}{4\nu t}\right), \quad P = 0, \tag{14.40}$$

where α and β are some arbitrary constants. This is the so called N-waves solution of the dissipative Burgers equation [105]. Substituting (14.40) into (14.38), we get

$$A[-1] = -\frac{2\nu}{\xi} + \frac{\alpha\xi}{tf}, \quad P[-1] = -\frac{\nu\xi(f-\alpha)}{2tf}.$$

It easy to see that if $t \to +\infty$ then

$$P[-1] \to 0, \quad A[-1] \to A_B = -\frac{2\nu}{\xi} + \frac{\xi}{t},$$

and A_B is the solution of the Burgers equation.

Chapter 15

Goursat Equation

The Goursat equation (GE) has the form [39]:

$$\zeta_{xy} = 2\sqrt{\lambda\,\zeta_x\zeta_y}, \tag{15.1}$$

where $\zeta = \zeta(x,y)$, $\lambda = \lambda(x,y)$. We call λ a potential function. This equation can be linearized by the substitution: $\psi = \sqrt{\zeta_x}$, and $\chi = \sqrt{\zeta_y}$. We have

$$\psi_y = \sqrt{\lambda}\chi, \quad \chi_x = \sqrt{\lambda}\psi$$

or

$$\psi_{xy} = \frac{1}{2}(\ln\lambda)_x\psi_y + \lambda\psi, \tag{15.2}$$

and a similar equation for χ. Equation (15.2) is a particular case of the following equation studied in the previous chapter,

$$\psi_{xy} + a\psi_y + b\psi = 0, \tag{15.3}$$

which has the Laplace transformations (LT)

$$a \to a_{-1} = a - \partial_x \ln(b - a_y), \quad b \to b_{-1} = b - a_y, \quad \psi \to \psi_{-1} = \psi_x + a\psi, \tag{15.4}$$

$$a \to a_1 = a + \partial_x \ln b, \quad b \to b_1 = b + \partial_y\left(a + \partial_x \ln b\right), \quad \psi \to \psi_1 = \frac{\psi_y}{b}; \tag{15.5}$$

and the Darboux transformations (DT)

$$a \to a_1 = a - \partial_x \ln(a + \tau), \quad b \to b_1 = b + \tau_y, \quad \psi \to \psi_1 = \psi_x - \tau\psi, \tag{15.6}$$

$$a \to {}_1a = -(\tau + b\rho), \quad b \to {}_1b = b - (b\rho)_y, \quad \psi \to {}_1\psi = \rho\psi_y - \psi. \tag{15.7}$$

where $\tau = \phi_x/\phi$, $\rho = \phi/\phi_y$, ψ and ϕ are particular solutions of (15.3).

The aim of this chapter is to study the validity of LT and DT for the GE. It is clear that after single DT or LT, the reduction restriction

$$a = -\partial_x \ln b, \tag{15.8}$$

will be true only for a special class of potentials. GE has two major applications:

(1) Let x be a complex variable, $y=-\overline{x}$, $\sqrt{\lambda}$ is a real-valued function, and ψ and χ in (15.3) are complex-valued functions. Then one defines three real-valued functions X_i, $i=1,2,3$ which are the coordinates of a surface in $\mathbb{R}^3$ [52]:

$$\begin{aligned} X_1+iX_2 &= 2i\int_\Gamma \left(\overline{\psi^2}dy'-\overline{\chi^2}dx'\right), \\ X_1-iX_2 &= -2i\int_\Gamma \left(\psi^2dy'-\chi^2dx'\right), \\ X_3 &= -2\int_\Gamma \left(\overline{\psi}\chi dy'+\overline{\chi}\psi dx'\right), \end{aligned} \tag{15.9}$$

where Γ is an arbitrary path of integration in the complex plane. The corresponding first fundamental form, the Gaussian curvature K and the mean curvature H yield:

$$ds^2=4U^2dxdy, \quad K=\frac{1}{U^2}\partial_x\partial_y\ln U, \quad H=\frac{\sqrt{\lambda}}{U},$$

where

$$U=|\psi|^2+|\chi|^2,$$

and any analytic surface in $\mathbb{R}^3$ can be globally represented by (15.9) (see [54]).

(2) The 2D-MKdV equation has the form

$$\begin{aligned} &4\lambda^2(\lambda_t-A\lambda_x+B\lambda_y-\lambda_{3x}-\lambda_{3y})+4\lambda^3\left[(2\lambda+B)_y+(2\lambda-A)_x\right] \\ &+6\lambda(\lambda_y\lambda_{yy}+\lambda_x\lambda_{xx})-3(\lambda_x^3+\lambda_y^3)=0, \\ &B_x=3\lambda_y-\lambda_x, \quad A_y=\lambda_y-3\lambda_x; \end{aligned} \tag{15.10}$$

where $\lambda=\lambda(x,y,t)$, $A=A(x,y,t)$, $B=B(x,y,t)$. If we introduce the function $u=\sqrt{\lambda}$, then we can rewrite (15.10) in the more customary form (see [21]):

$$\begin{aligned} &u_t+2u^2(u_x+u_y)+\frac{1}{2}(B_y-A_x)u+Bu_y-Au_x-u_{3y}-u_{3x}=0 \\ &B_x=(3\partial_y-\partial_x)u^2, \quad A_y=(\partial_y-3\partial_x)u^2. \end{aligned} \tag{15.11}$$

The reduction conditions

$$A=-B=-2u^2, \quad u_y=u_x,$$

lead to the MKdV equation,

$$u_t+12u^2u_x-2u_{3x}.$$

The 2D-MKdV equation (15.10) has a Lax pair which is given by (15.2) and

$$\psi_t=\psi_{3x}+\psi_{3y}-\frac{3}{2}\frac{\lambda_y}{\lambda}\psi_{yy}+\left[\frac{3}{4}\left(\frac{\lambda_y}{\lambda}\right)^2-\lambda-B\right]\psi_y+(A-\lambda)\psi_x+\frac{1}{2}(A_x-\lambda_x)\psi.$$

15.1 The Reduction Restriction

The reduction restriction (15.8) is valid only for a special type of potentials. The invariance of the reduction restriction under the Laplace transformations (15.4) is guaranteed by

$$\lambda_{-1} = \lambda - \frac{1}{2}\partial_x\partial_y \ln \lambda = \frac{C}{2\lambda}, \tag{15.12}$$

which is the well-known sinh-Gordon equation

$$\partial_x\partial_y \ln \lambda = 2\lambda - \frac{C}{\lambda}, \tag{15.13}$$

where C is a constant, and the new potential λ_{-1} is a solution of (15.13) too. In this case, the GE may be integrated and

$$\lambda = \frac{f'g'}{(f+g)^2}, \quad \zeta = -\frac{1}{C^2}\partial_y \ln(f+g) + V,$$

where $f = f(x)$ and $g = g(y)$ are arbitrary differentiable functions, and $V = V(y)$ is the function such that

$$V' = \left[\frac{1}{2C}\left(\ln g'\right)'\right]^2 = \frac{1}{4C^2}\left(\frac{g''}{g'}\right)^2,$$

and

$$\psi = \frac{\sqrt{f'g'}}{C(f+g)}, \quad \chi = \frac{1}{2C}\partial_y \ln\left(-\partial_y \frac{1}{f+g}\right).$$

Next we consider the DT (15.6). Inserting both transforms into the reduction condition (15.8) yields

$$\lambda_1 = \lambda - \tau_y = \lambda\left(\tau - \frac{\lambda_x}{2\lambda}\right). \tag{15.14}$$

Using the new notations $\alpha = \ln\phi$, $\Lambda = \ln\lambda$, and noticing that

$$\lambda - \tau_y = \left(-\frac{1}{2}\Lambda_x + \alpha_x\right)\alpha_y,$$

and $\tau = \alpha_x$, one gets from the transform (15.14) the condition for Λ:

$$\left(\alpha_x - \frac{1}{2}\Lambda_x\right)\left[\alpha_y - \exp(\Lambda)\left(\alpha_x - \frac{1}{2}\Lambda_x\right)\right] = 0. \tag{15.15}$$

Setting the first parentheses to zero yields:

$$\Lambda_{xy} = 2\exp(\Lambda),$$

and $\alpha = \Lambda/2 - c(y)$, where $c(y)$ is a arbitrary function. Setting the square brackets in (15.15) to zero yields: relevant equation

$$(\exp(-2\alpha)\lambda)_x = (\exp(-2\alpha))_y. \tag{15.16}$$

Thus

$$\theta_x = \psi^2 = \frac{1}{F_x + C_2}, \qquad \lambda = \frac{F_y + C_1}{F_x + C_2},$$

where $F = F(x,y)$ is any differentiable function, and C_1, C_2 are constants. Substituting (15.16) into (15.2), we get

$$\begin{aligned}&2(C_2 + F_x)C_1^2 + [(F_{yxx} + 4F_y)C_2 + F_x F_{yxx} + 4F_y F_x - F_{xx}F_{yx}]\, C_1\\&+ \left(F_{yxx}F_y - \frac{1}{2}F_{yx}^2 + 2F_y^2\right) C_2 + 2F_y^2 F_x - \frac{1}{2}F_{yx}^2 F_x\\&- F_y F_{xx} F_{yx} + F_x F_y F_{yxx} = 0.\end{aligned} \tag{15.17}$$

Introducing the new variables

$$F_x = P - C_2, \quad F_y = Q - C_1,$$

then (15.17) can be splitted into the system

$$2Q_x Q P_x - (2Q_{xx}Q - Q_x^2 + 4Q^2)P = 0, \quad P_y = Q_x. \tag{15.18}$$

After an integration of the first equation, we get

$$P = \frac{CQ_x}{\sqrt{Q}} \exp{(G)}, \quad G_x = 2\frac{Q}{Q_x},$$

where C is the third constant of integration. Let

$$Q = n^2, \quad G = \ln m,$$

where $m = m(x,y)$ and $n = n(x,y)$. The reduction equation then takes the simple form

$$\left(n^2\right)_x = 2C\left(mn_x\right)_y, \quad m_x n_x = mn. \tag{15.19}$$

This system can be rewritten in a more convenient form. Let

$$n_x = n\exp{(S)}, \quad m_x = m\exp{(-S)},$$

where $S = S(x,y)$. After substituting into (15.19), we get

$$S_y = \frac{1}{C}\frac{n}{m} - \partial_y \ln(mn),$$

therefore

$$S_{xy} = 4\sinh S\, \partial_y \partial_x^{-1} \cosh S, \tag{15.20}$$

which is a 2D generalization of the sinh-Gordon equation. Equation (15.20) has the Lax pair:

$$K\psi = 0, \quad K_1 D\psi = 0$$

where

$$K = \partial_x\partial_y - \frac{1}{2}\frac{\lambda_x}{\lambda}\partial_y - \lambda,\ K_1 = \partial_x\partial_y - \frac{1}{2}\frac{\lambda_{1,x}}{\lambda_1}\partial_y - \lambda_1,\ D = \partial_x - \tau,$$

the variables λ and λ_1 are defined by

$$\lambda = \frac{(S_x + 2\cosh S)_y}{4\sinh S}\exp(-S), \quad \lambda_1 = \frac{(S_x + 2\cosh S)_y}{4\sinh S}\exp(S), \tag{15.21}$$

and

$$\tau_y \equiv \lambda - \lambda_1.$$

We can study the reduction for the DT (15.7) analogously. As a result we get

$$\lambda = C_1\phi_y\exp(F), \quad {}_1\lambda = -\frac{C_1C_2\phi^2}{\phi_y}\exp(F), \tag{15.22}$$

where ϕ is the support function of the DT (15.7) and the reduction equation can be written as

$$\phi_{xy} = \phi_y[F_x + 2C_1\phi\exp(F)], \quad F_y\phi_y = C_2\phi.$$

15.2 Binary Darboux Transformations

In [36], Ganzha studied the one analogue of the Moutard transformation for the Goursat equation, which is valid without a reduction restriction. Here we will present a binary Darboux transformation for the GE with the same property.

We introduce the new variables ξ and η,

$$\partial_y = \partial_\eta - \partial_\xi, \quad \partial_x = \partial_\eta + \partial_\xi,$$

and rewrite (15.2) in the matrix form

$$\Psi_\eta = \sigma_3\Psi_\xi + U\Psi, \tag{15.23}$$

where

$$\Psi = \begin{pmatrix} \psi_1 & \psi_2 \\ \chi_1 & \chi_2 \end{pmatrix}, \quad U = \sqrt{\lambda}\sigma_1, \tag{15.24}$$

where $\psi_k = \psi_k(\xi, \eta)$ and $\chi_k = \chi_k(\xi, \eta)$ $(k = 1, 2)$ are particular solutions of (15.3) for some $\lambda(\xi, \eta)$, and σ_1, σ_3 are the Pauli matrices. Let Ψ_1 be a solution of the equation (15.23), $\Psi \neq \Psi_1$; we define a matrix function $\tau = \Psi_{1,\xi}\Psi_1^{-1}$. The equation (15.23) is covariant with respect to the DT

$$\Phi[1] = \Phi_\xi - \tau\Phi, \quad U[1] = U + [\sigma_3, \tau]. \tag{15.25}$$

Remark 15.1. It is not difficult to check that the DT (15.25) is the superposition formula for the two simpler Darboux transformations given by formulas (15.6) and (15.7). Eq. (15.23) is the spectral problem for the Davey-Stewartson (DS) equations.

Let us consider a closed 1-form

$$d\Omega = d\xi\, \Phi\Psi + d\eta\, \Phi\sigma_3\Psi, \quad \Omega \equiv \int d\Omega,$$

where a 2×2 matrix function Φ solves the equation:

$$\Phi_\eta = \Phi_\xi\sigma_3 - \Phi U. \tag{15.26}$$

One can verify that (15.26) is covariant with respect to the transform (15.25) if

$$\Phi[+1] = \Omega(\Phi, \Psi_1)\Psi_1^{-1}.$$

Now we can transform U by

$$U[+1,-1] = U + [\sigma_3, \Psi_1\Omega^{-1}\Phi].$$

A particular solution of equation (15.26) has the form

$$\Phi_1 = \begin{pmatrix} s_1\psi_1 + s_2\psi_2 & -s_1\chi_1 - s_2\chi_2 \\ s_3\psi_1 + s_4\psi_2 & -s_3\chi_1 - s_4\chi_2 \end{pmatrix}, \tag{15.27}$$

where s_k $(k = 1, \ldots, 4)$ are constants. It is convenient to choose

$$\Phi_1 = \Psi_1^T\sigma_3, \tag{15.28}$$

where Ψ_1^T is the transposed matrix Ψ_1. In this case

$$U[+1,-1] = U - 2A_{_F}, \tag{15.29}$$

where $A_{_F}$ is the off-diagonal part of the matrix $A = \Psi_1\Omega^{-1}\Psi_1^T$, $\Omega = \Omega(\Phi_1, \Psi_1)$, and

$$A_{_F}^T = A_{_F} = f\sigma_1 \tag{15.30}$$

where $f = f(\xi, \eta)$ is an arbitrary function. Using (15.24), (15.29) and (15.30), we can see that $U[+1,-1]$ has the same form as the initial matrix U:

$$U[+1,-1] \equiv \begin{pmatrix} 0 & \sqrt{\lambda[+1,-1]} \\ \sqrt{\lambda[+1,-1]} & 0 \end{pmatrix} = \begin{pmatrix} 0 & \sqrt{\lambda} - 2f \\ \sqrt{\lambda} - 2f & 0 \end{pmatrix},$$

thus the reduction restriction is valid. The new function $\Phi[+1,-1]$ has the form

$$\Phi[+1,-1] = \Phi - \Omega(\Phi, \Psi_1)\left(\Omega(\Phi_1, \Psi_1)\right)^{-1}\Phi_1, \tag{15.31}$$

where Φ is an arbitrary solution of equation (15.26).

Theorem 15.2. *Let*

$$\psi_{k,y} = \sqrt{\lambda}\chi_k, \quad \chi_{k,x} = \sqrt{\lambda}\psi_k,$$
$$\alpha_{k,y} = -\sqrt{\lambda}\beta_k, \quad \beta_{k,x} = -\sqrt{\lambda}\alpha_k,$$

where $k = 1, 2$; then the new functions:

$$\alpha_1' = \alpha_1 - \frac{A_1\psi_1 + A_2\psi_2}{D}, \quad \beta_1' = \beta_1 + \frac{A_1\chi_1 + A_2\chi_2}{D},$$

are solutions of the equations

$$\alpha_{1,y}' = \sqrt{\lambda'}\,\beta_1', \quad \beta_{1,x}' = \sqrt{\lambda'}\,\alpha_1',$$

where

$$\sqrt{\lambda'} = -\sqrt{\lambda} + \frac{\psi_1\chi_1\Omega_{22} + \psi_2\chi_2\Omega_{11} - (\psi_1\chi_2 + \psi_2\chi_1)\Omega_{12}}{D},$$

and

$$\Omega_{11} = \int dx\psi_1^2 + dy\chi_1^2, \quad \Omega_{12} = \Omega_{21} = \int dx\psi_1\psi_2 + dy\chi_1\chi_2,$$
$$\Omega_{22} = \int dx\psi_2^2 + dy\chi_2^2, \quad D = \Omega_{11}\Omega_{22} - \Omega_{12}^2,$$
$$\Lambda_{11} = \int dx\alpha_1\psi_1 + dy\beta_1\chi_1, \quad \Lambda_{12} = \int dx\alpha_1\psi_2 + dy\beta_1\chi_2,$$
$$\Lambda_{21} = \int dx\alpha_2\psi_1 + dy\beta_2\chi_1, \quad \Lambda_{22} = \int dx\alpha_2\psi_2 + \beta_2\chi_2.$$
$$A_1 = \Lambda_{11}\Omega_{22} - \Lambda_{12}\Omega_{12}, \quad A_2 = \Lambda_{12}\Omega_{11} - \Lambda_{11}\Omega_{12},$$

where $\int = \int_\Gamma$, and Γ is an arbitrary path of integration on the plane.

15.3 Moutard Transformations for 2D-MKdV Equation

Recall that the Lax pair of the 2D-MKdV equation (15.11) has the form,

$$\begin{aligned}\psi_{xy} &= \frac{u_x}{u}\psi_y + u^2\psi,\\ \psi_t &= \psi_{3x} + \psi_{3y} - 3\frac{u_y}{u}\psi_{yy} + \left[3\left(\frac{u_y}{u}\right)^2 - u^2 - B\right]\psi_y\\ &\quad + \left(A - u^2\right)\psi_x + \frac{1}{2}\left(A - u^2\right)_x\psi.\end{aligned} \tag{15.32}$$

Let ϕ be a solution of (15.32) (the support function), then we have a closed 1-form,

$$d\theta = dx\,\theta_1 + dy\,\theta_2 + dt\,\theta_3, \quad \theta = \int d\theta,$$

where

$$\theta_1 = \phi^2, \quad \theta_2 = \left(\frac{\phi_y}{u}\right)^2,$$
$$\theta_3 = (A - u^2)\phi^2 - \phi_y^2 - \phi_x^2 + 2\phi\phi_{xx} + \\ + \frac{(2\phi_{3y}\phi_y - \phi_{yy}^2 - B\phi_y^2)u^2 - 2u\phi_y(u_y\phi_y)_y + 3\,(u_y\phi_y)^2}{u^4}.$$

We define the generalized Moutard transformation by

$$u \to \tilde{u} = u - \sqrt{(\ln\theta)_x(\ln\theta)_y}, \quad A \to \tilde{A} = A - (\partial_x\partial_y - 3\partial_x^2)\ln\theta,$$
$$B \to \tilde{B} = B + (\partial_x\partial_y - 3\partial_y^2)\ln\theta, \quad \psi \to \tilde{\psi} = \frac{\phi Q}{\theta}, \tag{15.33}$$

where

$$Q \equiv \int dQ, \quad dQ = dx\,Q_1 + dy\,Q_2 + dt\,Q_3,$$

and $(w = \psi/\phi)$,

$$Q_1 = \theta w_x, \quad Q_2 = -\frac{\theta^3(1/\theta)_{xy}w_y}{\theta_{xy}},$$
$$Q_3 = \theta w_{3x} + c_1 w_{3y} + c_2 w_{xx} + c_3 w_{yy} + c_4 w_x + c_5 w_y,$$

with

$$c_1 = -\frac{\theta_{xy}}{2u^2} + \theta, \quad c_2 = \frac{3}{2}\theta(\ln\theta_x)_x - \theta_x, \quad c_4 = \left(\frac{3\phi_{xx}}{\phi} + A - u^2\right)\theta - \frac{\theta_{xx}}{2},$$
$$c_3 = \frac{u_y\theta_{xy}}{2u^3} + \frac{\phi\phi_{yy}}{u^2} - \frac{3u_y\theta}{u} + 3\left(\frac{\theta}{2}(\ln\theta_x)_y - \theta_y\right),$$
$$c_5 = -\frac{3u_y^2\theta_{xy}}{2u^4} + \frac{1}{u^3}(\theta_{xy}u_{yy} + u_y\phi\phi_{yy}) + \frac{1}{u^2}\left[3\theta u_y^2 - \phi\phi_{3y} + \frac{1}{2}\left(B - \frac{\phi_{yy}}{\phi}\right)\theta_{xy}\right] \\ + \left(\frac{3\phi_{yy}}{\phi} - B\right)\theta + \frac{u_y}{u}\left(2\theta_y - \frac{3\theta\theta_{xy}}{\theta_x}\right) + \frac{\theta_{xy}}{2} - u^2\theta.$$

The 1-form dQ is closed,

$$Q_{1,y} = Q_{2,x}, \quad Q_{1,t} = Q_{3,x}, \quad Q_{2,t} = Q_{3,y}.$$

It is easy to verify that the Lax pair (15.32) is covariant with respect to the generalized Moutard transformation (15.33).

Now we use these transformations to construct exact solutions of the 2D-MKdV equation (15.11). Let u = const., $A = B = 0$. We will consider two examples.

(1) If we choose the solution of (15.32) as $\phi = \sinh\xi$, where

$$\xi = ax + \frac{u^2}{a}y + \frac{(u^2 - a^2)(u^4 - a^4)}{a^3}t, \tag{15.34}$$

with the real constant a, then using (15.33) we get new solutions of the 2D-MKdV equation,

$$\tilde{u} = \frac{u\left[2\eta - a^3 \sinh(2\xi)\right]}{2\eta + a^3 \sinh(2\xi)}, \quad \tilde{A} = \frac{16a^3 \sinh\xi \left[3a^5 \sinh\xi - (u^2 - 3a^2)\eta \cosh\xi\right]}{\left[2\eta + a^3 \sinh(2\xi)\right]^2},$$
$$\tilde{B} = \frac{16av^2 \cosh\xi \left[3a^3u^2 \cosh\xi - (3u^2 - a^2)\eta \sinh\xi\right]}{\left[2\eta + a^3 \sinh(2\xi)\right]^2},$$

where

$$\eta = a^2(u^2y - a^2x) + (u^2 - a^2)(3u^4 + 3a^4 + 2a^2u^2)t. \tag{15.35}$$

(2) To construct the algebraic solutions of (15.11), we choose the solution of (15.32) as,

$$\phi = (-1)^n \int_\alpha^\beta dk\zeta(k) \exp\xi(k) \frac{d^n}{dk^n}\delta(k - k_0),$$

with $\xi(k)$ given by (15.34), $a = a(k)$, $\beta > k_0 > \alpha > 0$, $\zeta(k)$ is an arbitrary differentiable function. Choosing $n = 1$, $\zeta = 1$, we get

$$\begin{aligned} \tilde{u} &= \frac{u(a^6 - 2\eta^2 - 2a^3\eta)}{2\eta^2 + 2a^3\eta + a^6}, \quad \tilde{A} = -\frac{8a^6(u^2 + 3a^2)\eta(\eta + a^3)}{(2\eta^2 + 2a^3\eta + a^6)^2}, \\ \tilde{B} &= \frac{8u^2a^4(3u^2 + a^2)\eta(\eta + a^3)}{(2\eta^2 + 2a^3\eta + a^6)^2}, \end{aligned} \tag{15.36}$$

with the η given by (15.35), and $a = a(k_0)$. (15.36) is the simple nonsingular algebraic solution of the 2D-MKdV equation.

Chapter 16

Links Among Integrable Systems

An interesting and rather philosophical question is whether or not there is a hidden link among all integrable systems. This is obviously a very difficult question which is still open. Partial progress was made via Darboux transformations. For instance, Korteweg-de Vries equation (KdV), sine-Gordon equation (SG), Calogero-Degasperis equation (CD), and higher order modified KdV are linked [22]. Also KdV, m^{n}KdV ($n = 1, 2, 3$), and Krichever-Novikov equation exhaust all the integrable equations of the form $u_t + u_{xxx} + f(u_{xx}, u_x, u) = 0$ [22, 88].

16.1 Borisov-Zykov's Method

Consider the KdV equation

$$u_t - 6uu_x + u_{xxx} = 0, \tag{16.1}$$

with its Lax pair

$$\psi_{xx} = (u - \lambda)\psi, \quad \psi_t = 2(u + 2\lambda)\psi_x - u_x\psi. \tag{16.2}$$

Setting $\tau = \phi_x/\phi$, where ϕ is a solution of (16.2) at $\lambda = \mu$, then

$$\tau_x = -\tau^2 + u - \mu, \quad \tau_t = [2(u + 2\mu)\tau - u_x]_x \,. \tag{16.3}$$

Eliminating u from (16.3), we get

$$\tau_t = 6\tau^2\tau_x - \tau_{xxx} + 6\mu\tau_x. \tag{16.4}$$

If $\mu = 0$, then we have the well-known mKdV equation. To construct the Lax pair for mKdV, we use the DT

$$u_1 = u - 2\tau_x, \quad \psi_1 = \psi_x - \tau\psi.$$

Setting $\sigma = \psi_{1,x}/\psi_1$, we get the x-chain

$$(\sigma + \tau)_x = -\sigma^2 + \tau^2 - \lambda + \mu. \tag{16.5}$$

and the t-chain,

$$(\sigma + \tau)_t = \left[2(-\tau_x + \tau^2 + \mu + 2\lambda)\sigma - 2\tau\tau_x + 6\mu\tau + 2\tau^3\right]_x.$$

Let

$$\sigma + \tau = \Psi,$$

we obtain the Lax pair for mKdV (16.4),

$$\Psi_x = -\Psi^2 + 2\tau\Psi - \lambda + \mu, \quad \Psi_t = 2\left[(\tau^2 - \tau_x + \mu + 2\lambda)\Psi + 2(\mu - \lambda)\tau\right]_x. \tag{16.6}$$

Continuing the above procedure, one can find higher order modified KdV, m^2KdV and m^3KdV. m^2KdV is the so-called Calogero-Degasperis equation (CD). For m^3KdV, one can not rewrite its Lax pair as a pair of Riccati equations anymore. So the above procedure stops at m^3KdV. Nevertheless, as mentioned above, KdV, m^nKdV ($n = 1, 2, 3$), and Krichever-Novikov equation exhaust all the integrable equations of the form $u_t + u_{xxx} + f(u_{xx}, u_x, u) = 0$ [22, 88].

In [6], the lower order equations in the KdV hierarchy were constructed. Let

$$L = -\partial_x^2 + u(x,t),$$

and the nonlinear equation be given by

$$L_t = [L, A_N].$$

where the operators A_N have the form

$$A_N = \sum_{m=-1}^{N} K_m (L)^m,$$

where $N = -1, -2, \ldots$ and K_m are some operators. The first lower order KdV ($N = -1$) equation has the form,

$$\text{KdV}_{-1}(\sigma) \equiv (\sigma_x^2 + \sigma_{xx})_t - \left(e^{2\sigma}\right)_x = 0,$$

where σ is connected with u by

$$u = -\sigma_{xx} - \sigma_x^2.$$

Let $\sigma = iq/2$, then

$$\text{KdV}_{-1}(\sigma) = \frac{1}{2}\left(i\partial_x - q_x\right)\left(q_{xt} - 2\sin q\right) = 0. \tag{16.7}$$

Thus we have a Miura transformation ($u \to q$) between the KdV$_{-1}$ equation and the sine-Gordon equation. The Lax pair of (16.7) has the form,

$$\psi_{xx} = \left(\frac{iq_{xx}}{2} - \frac{q_x^2}{4} + \lambda^2\right)\psi, \quad \psi_t = \frac{1}{2\lambda^2}e^{iq}\left(\psi_x - \frac{iq_x}{2}\psi\right). \tag{16.8}$$

Starting out from (16.8), we can find the Lax pair for the sine-Gordon equation, by introducing the new function $\tilde{\psi}$

$$\tilde{\psi} = \frac{1}{\lambda}\left(\psi_x - \frac{i}{2}q_x\psi\right). \tag{16.9}$$

Then

$$\tilde{\psi}_x = \lambda\psi - \frac{i}{2}q_x\tilde{\psi}, \quad \psi_t = \frac{1}{2\lambda}e^{iq}\tilde{\psi}, \quad \tilde{\psi}_t = -\frac{1}{2\lambda}\left(iq_{xt} - e^{iq}\right)\psi. \tag{16.10}$$

Replacing q_{xt} by $2\sin q$ in the equation for $\tilde{\psi}_t$, we see that (16.9) and (16.10) are the well known Lax pair for the sine-Gordon equation.

16.2 Higher Dimensional Systems

The Borisov-Zykov's method can be applied to higher dimensional systems too. Here we consider the Kadomtsev-Petviashvili (KP) equation

$$u_t + 6uu_x + u_{xxx} = -3\alpha^2 v_y, \quad v_x = u_y, \tag{16.11}$$

where $u = u(x,y,t)$, $v = v(x,y,t)$, and $\alpha^2 = \pm 1$. The Lax pair for KP is

$$\alpha\psi_y + \psi_{xx} + u\psi = 0, \quad \psi_t + 4\psi_{xxx} + 6u\psi_x + 3(u_x - \alpha v)\psi = 0. \tag{16.12}$$

Setting $f = \ln\phi$, $\tau = f_x$, where ϕ is a solution of (16.12), we can rewrite the Lax pair (16.12) as

$$\begin{aligned} \alpha\tau_y + \left(\tau_x + \tau^2 + u\right)_x &= 0, \\ \tau_t + \left(4\tau_{xx} + 12\tau\tau_x + 4\tau^3 + 6u\tau + 3u_x - 3\alpha v\right)_x &= 0. \end{aligned} \tag{16.13}$$

It follows from the first equation of (16.13) that

$$u = -\tau_x - \tau^2 - \alpha F, \quad F_x = \tau_y. \tag{16.14}$$

Substituting (16.14) into the second equation in (16.13), we obtain the well-known mKP equation [53]

$$\tau_t - 6\tau^2\tau_x + \tau_{xxx} = 3\alpha(2\tau_x F - \alpha F_y), \quad F_x - \tau_y = 0. \tag{16.15}$$

The KP equation (16.11) admits the Darboux transformation [84]

$$\psi_1 = \psi - \tau\psi, \quad u_1 = u + 2\tau_x, \quad v_1 = v + 2\tau_y. \tag{16.16}$$

Setting $s = \ln\psi_1$, $\sigma = s_x$, we see that σ satisfies the system of equations obtained from (16.13) by the replacements $u \to u_1$ and $v \to v_1$. Comparing these equations with (16.13) and eliminating the potentials u and v, we obtain

$$\begin{aligned} &\alpha(s-f)_y + (s+f)_{xx} + s_x^2 - f_x^2 = 0, \\ &(s-f)_t + \left[2(2s+f)_{xx} + 6s_x^2 - 3f_x^2\right]_x + 4s_x^3 \\ &+6\left(f_{xx} - f_x^2 - \alpha f_y\right)s_x - 6\alpha(f_{xy} - f_x f_y) + 2f_x^3 = 0. \end{aligned} \tag{16.17}$$

Let $\Psi = s - f$, we obtain

$$\begin{aligned} &\alpha\Psi_y + (2f+\Psi)_{xx} + 2f_x\Psi_x + \Psi_x^2 = 0, \\ &\Psi_t + 2\left(2\Psi_{xx} + 6f_x\Psi_x + 3\Psi_x^2 + 3\theta\right)_x + 6\theta\Psi_x + 4\left(3f+\Psi\right)_x\Psi_x^2 = 0, \\ &\theta \equiv f_{xx} + f_x^2 - \alpha f_y. \end{aligned} \tag{16.18}$$

The compatibility condition for these equations is

$$\left(f_t - 2f_x^3 + f_{xxx}\right)_x = 3\alpha\left(2f_{xx}f_y - \alpha f_{yy}\right),$$

which is merely another form of the mKP equation (with $\tau = f_x$ and $F = f_y$).

This process can be continued. In particular, the system (16.18) contains a new nonlinear equation obtained by eliminating the potential f. Let

$$f = -\frac{1}{2}\left(\Psi + \alpha\xi\right), \tag{16.19}$$

where $\xi = \xi(x,y,t)$. Substituting (16.19) into the second equation of (16.18), differentiating it with respect to x, and introducing $S = \Psi_x$, we obtain the system

$$\begin{aligned} &S_t + S_{xxx} - \frac{3}{2}S^2S_x = -3\alpha^2\left[\left(\frac{1}{2}\xi_x^2 + \xi_y\right)S + \left(\frac{1}{2}\xi_x^2 + \xi_y\right)_x\right]_x, \\ &S_y = \left(\xi_{xx} + \xi_x S\right)_x. \end{aligned} \tag{16.20}$$

Although the system (16.20) looks like a two-dimensional generalization of the mKdV equation, its one-dimensional limit ($\partial_y = 0$)

$$g_t + \left(g_{xx} - \frac{\alpha^2}{2}g^3 - \frac{3}{2}\frac{g_x^2}{g}\right)_x = 0,$$

(where $g(x,t) = \xi_x = \exp(-\Psi)$) reduces to the exponential Calogero-Degasperis equation (CD).

16.3 Modified Nonlinear Schrödinger Equations

Consider the nonlinear Schrödinger equations (NLS)

$$iu_t + u_{xx} + 2u^2v = 0, \quad -iv_t + v_{xx} + 2v^2u = 0. \tag{16.21}$$

It has the Lax pair

$$\Psi_t = -2i\sigma_3\Psi\Lambda^2 + 2iU\Psi\Lambda + V\Psi, \quad \Psi_x = -i\sigma_3\Psi\Lambda + iU\Psi, \tag{16.22}$$

where

$$U = \begin{pmatrix} 0 & u \\ v & 0 \end{pmatrix}, \quad \sigma_3 = \begin{pmatrix} 1 & 0 \\ 0 & -1 \end{pmatrix}, \quad V = \sigma_3\left(iU^2 - U_x\right),$$

and Λ is arbitrary constant matrix.

Let Φ and Ψ be solutions of (16.22) with

$$\Lambda = \mathrm{diag}\left((\lambda+\mu)/2, (\lambda-\mu)/2\right), \quad \text{and } \Lambda_1 = \mathrm{diag}\left((\lambda_1+\mu_1)/2, (\lambda_1-\mu_1)/2\right),$$

respectively, and $\tau = \Phi\Lambda\Phi^{-1}$. Then

$$\tau_x = i[\tau, \sigma_3]\tau + i[U, \tau], \quad \tau_t = 2\tau_x\tau + [V, \tau]. \tag{16.23}$$

The Darboux transformation has the form

$$\Psi \to \Psi_1 = \Psi\Lambda_1 - \tau\Psi, \quad U \to U_1 = U + [\tau, \sigma_3]. \tag{16.24}$$

Thus

$$u \to u_1 = u - 2b, \quad v \to v_1 = v + 2c, \tag{16.25}$$

where

$$\tau = \begin{pmatrix} a & b \\ c & d \end{pmatrix}. \tag{16.26}$$

Substituting (16.26) into (16.23), we get

$$\begin{aligned} a_x &= -d_x = -i(2bc - uc + vb), \\ b_x &= -i(2bd + (a-d)u), \\ c_x &= i(2ac + (a-d)v), \end{aligned} \tag{16.27}$$

and

$$\begin{aligned} a_t &= (a^2)_x + 2b_xc - bv_x - cu_x, \\ d_t &= (d^2)_x + 2bc_x + bv_x + cu_x, \\ b_t &= 2(ibuv + a_xb + b_xd) + (a-d)u_x, \\ c_t &= 2(-icuv + ac_x + cd_x) + (a-d)v_x. \end{aligned} \tag{16.28}$$

Calculating the determinant and trace of matrix τ we get

$$ad - bc = \frac{\lambda^2 - \mu^2}{4}, \quad a + d = \lambda.$$

Eliminating u and v from the last two equations in (16.27), we get the mNLS equation

$$\begin{aligned}(\mu^2 - 4bc)\left(ib_t + b_{xx} - 2b^2c\right) + 2\lambda\left(\lambda c + 2ic_x\right)b^2 + 2(b_xc + 2bc_x)b_x = 0,\\ (\mu^2 - 4bc)\left(-ic_t + c_{xx} - 2c^2b\right) + 2\lambda\left(\lambda b - 2ib_x\right)c^2 + 2(bc_x + 2b_xc)c_x = 0.\end{aligned}$$

16.4 NLS and Toda Lattice

In the Lax pair (16.22) of NLS, let

$$\Lambda = \begin{pmatrix} \lambda & 0 \\ 0 & \mu \end{pmatrix},$$

where λ and μ are spectral parameters. Denote the components of the first column of the matrix Ψ by ψ_n and ϕ_n. Let $u = u_n$ and $v = v_n$. Then, the spatial part of (16.22) has the form

$$\psi_{n,x} = -i\lambda\psi_n + iu_n\phi_n, \quad \phi_{n,x} = i\lambda\phi_n + iv_n\psi_n. \tag{16.29}$$

The Schlesinger transformation of the NLS is given by

$$u_n \to u_{n+1} = u_n\left(u_nv_v + (\log u_n)_{xx}\right), \quad v_n \to v_{n+1} = \frac{1}{u_n}, \tag{16.30}$$

$$\psi_n \to \psi_{n+1} = \left(-2\lambda + i(\log u_n)_x\right)\psi_n + u_n\phi_n, \quad \phi_n \to \phi_{n+1} = \frac{\psi_n}{u_n}, \tag{16.31}$$

and its inverse is given by

$$v_n \to v_{n-1} = v_n\left(u_nv_v + (\log v_n)_{xx}\right), \quad u_n \to u_{n-1} = \frac{1}{v_n}, \tag{16.32}$$

$$\phi_n \to \phi_{n-1} = \left(2\lambda + i(\log v_n)_x\right)\phi_n + v_n\psi_n, \quad \psi_n \to \psi_{n-1} = \frac{\phi_n}{v_n}. \tag{16.33}$$

Denote by

$$q_n = \log u_n, \quad p_n = q_{n,x}, \quad U_n = \frac{u_n}{u_{n-1}} = e^{q_n - q_{n-1}}. \tag{16.34}$$

Then q_n satisfies the Toda lattice (TL) equation [4, 101]

$$q_{n,xx} = e^{q_{n+1} - q_n} - e^{q_n - q_{n-1}}. \tag{16.35}$$

One can also derive the Lax pair and Darboux transformations for the Toda lattice from those of NLS. Introduce the shift-operator T,

$$Tu_n = u_{n+1}.$$

Let T^{-1} act on the second equation of (16.31), then substitute the expression for ϕ_n into the first equation of (16.29), we obtain the first equation of the Lax pair for TL

$$\psi_{n,x} = -i\lambda\psi_n + iU_n\psi_{n-1}. \tag{16.36}$$

To obtain the second equation of the Lax pair, let T act on the second equation of (16.29), together with the expressions of ϕ_{n+1} from (16.31), v_{n+1} from (16.30), and $\psi_{n,x}$ from (16.36), we get the second equation of the Lax pair for TL

$$\psi_{n+1} = (2\lambda + ip_n)\psi_n + U_n\psi_{n-1}. \tag{16.37}$$

Let ψ_1 and ϕ_1 be the components of the first column of the matrix Ψ, where Ψ is a solution of (16.22) with $\lambda = \lambda_1$. Then one can write two DTs for NLS (the indices are omitted):

$$\begin{aligned} &\psi \to \psi^{(1)} = \left[2(\lambda - \lambda_1) + \frac{\phi_1}{\psi_1}u\right]\psi - u\phi, \quad \phi \to \phi^{(1)} = \phi - \frac{\phi_1}{\psi_1}\psi, \\ &u \to u^{(1)} = iu_x - \left(2\lambda_1 - \frac{\phi_1}{\psi_1}u\right)u, \quad v \to v^{(1)} = \frac{\phi_1}{\psi_1}, \end{aligned} \tag{16.38}$$

and

$$\begin{aligned} &\psi \to {}^{(1)}\psi = \psi - \frac{\psi_1}{\phi_1}\phi, \quad \phi \to {}^{(1)}\phi = \left[2(\lambda_1 - \lambda) + \frac{\psi_1}{\phi_1}v\right]\phi - v\psi, \\ &u \to {}^{(1)}u = \frac{\psi_1}{\phi_1}, \quad v \to {}^{(1)}v = iv_x + \left(2\lambda_1 + \frac{\psi_1}{\phi_1}v\right)v. \end{aligned} \tag{16.39}$$

Let ψ_2 and ϕ_2 be the components of the second column of the matrix Ψ with $\mu = \lambda_2$. Then one can iterate the Darboux transformations (16.38) and (16.39) to obtain

$$\begin{aligned} &u \to u^{(1)} \to {}^{(2)}u^{(1)}, \quad v \to v^{(1)} \to {}^{(2)}v^{(1)}, \\ &\psi \to \psi^{(1)} \to {}^{(2)}\psi^{(1)}, \quad \phi \to \phi^{(1)} \to {}^{(2)}\phi^{(1)}. \end{aligned} \tag{16.40}$$

Note that the DT transformations are commutative:

$${}^{(2)}u^{(1)} = {}^{(1)}u^{(2)}, \quad {}^{(2)}v^{(1)} = {}^{(1)}v^{(2)}.$$

Then it is easy to find the DT for TL. Let $\{\psi_{1,n}\}$ be the solution of the Lax pair (16.36) and (16.37) with $\lambda = \lambda_1$. Then we have two DTs for TL equations:

$$\psi_n \to \psi_n^{(1)} = -\psi_{n+1} + \frac{\psi_{1,n+1}}{\psi_{1,n}}\psi_n, \quad q_n \to q_n^{(1)} = q_n + \ln\frac{\psi_{1,n+1}}{\psi_{1,n}}; \tag{16.41}$$

and

$$\psi_n \to {}^{(1)}\psi_n = \psi_n - \frac{\psi_{1,n}}{\psi_{1,n-1}}\psi_{n-1}, \quad q_n \to {}^{(1)}q_n = q_{n-1} + \ln\frac{\psi_{1,n}}{\psi_{1,n-1}}. \tag{16.42}$$

Using (16.36) and (16.37), it is also easy to find the first modified Toda lattice equation. Define τ_n and ξ_n as follows

$$\tau_n = \frac{\psi_{1,n}}{\psi_{1,n+1}} = \frac{1}{\partial \xi_n},$$

where $\partial \equiv -i\partial/\partial x$, $V_n \equiv ip_n$. Then the Lax pair for TL can be rewritten as

$$\partial \tau_n = \tau_n \left(\tau_{n-1} U_n - \tau_n U_{n+1}\right), \quad \tau_{n+1} = \frac{1}{2\lambda_1 + V_{n+1} + \tau_n U_{n+1}}. \tag{16.43}$$

Eliminating the potentials U_n, U_{n+1} and V_{n+1}, one get the modified Toda lattice equation $\mathrm{m}^1\mathrm{TL}$, which is just a Volterra equation,

$$\partial^2 \xi_n = \partial \xi_n \left(e^{\xi_n - \xi_{n+1}} - e^{\xi_{n-1} - \xi_n}\right).$$

Bibliography

[1] M. Ablowitz and J. F. Ladik, A nonlinear difference scheme and inverse scattering, *Stud. Appl. Math.* **55** (1976), 213.

[2] M. Ablowitz, B. Prinari, and A. Trubatch, *Discrete and Continuous Nonlinear Schrödinger Systems*, London Mathematical Society Lecture Note Series, 2004.

[3] M. Ablowitz and H. Segur, *Solitons and the Inverse Scattering Transform*, SIAM, Philadelphia, 1981.

[4] V. A. Adler and A. B. Shabat, On the one class of the Toda chains, *Theor. Math. Phys.* **111, no.3** (1997), 323.

[5] R. L. Anderson and N. H. Ibragimov, *Lie-Bäcklund Transformations in Applications*, SIAM, Philadelphia, 1979.

[6] V. A. Andreev and M. V. Burova, Lower Korteweg-de Vries equations and supersymmetric structure of the sine-Gordon and Liouville equations, *Theor. Math. Phys.* **85** (1990), 376-387.

[7] A. Andrianov, N. Borisov, and M. Ioffe, The factorization method and quantum systems with equivalent energy spectra, *Phys. Lett. A* **105** (1984), 19.

[8] A. Andrianov, M. Borisov, M. Ioffe, and M. Eides, Supersymmetric origin of equivalent quantum systems, *Phys. Lett. A* **109** (1985), 143-148.

[9] A. Andrianov, M. Ioffe, and D. Nishnianidze, Polynomial supersymmetry and dynamical symmetries in quantum mechanics, *Theor. Math. Phys.* **104**, 463.

[10] V. E. Adler, A modification of Crum's method, *Theor. Math. Phys.* **101, no.3** (1994), 323-330.

[11] A. V. Bäcklund, Einiges uber curven- und flachen-transformationen, *Lunds Universitets Ars-skrift, Afdelningen for Mathematik och Naturetenskap* **X, II** (1873-1874), 1.

[12] A. V. Bäcklund, Ueber flachentransformationen, *Math. Ann.* **IX** (1876), 297.

[13] A. V. Bäcklund, Zur Theorie der partiellen differentialgleichungen erster ordnung, *Math. Ann.* **XVII** (1880), 285.

[14] A. V. Bäcklund, Zur Theorie der flachentransformationen, *Math. Ann.* **XIX** (1882), 387.

[15] A. V. Bäcklund, Om ytor med konstant negativ krokning, *Lunds Universitets Ars-skrift* **19** (1883).

[16] V. P. Berezovoy and A. I. Pashnev, Supersymmetric quantum mechanics and rearrangement of the spectra of Hamiltonians, *Theor. Math. Phys.* **70, no.1** (1987), 146-153.

[17] L. Berger, Emission of spin waves by a magnetic multilayer, *Phys. Rev. B* **54** (1996), 9353.

[18] L. Bianchi, Ricerche sulle superficie a curvatura constante e sulle elicoidi, *Ann. Scuola Norm. Sup. Pisa* **II** (1879), 285.

[19] L. Bianchi, *Lezioni di Geometria Differenziale*, volume I, Enrico Spoerri, Pisa, 1922.

[20] M. Boiti, J. Leon, and F. Pempinelli, Integrable two-dimensional generalisation of the sine- and sinh-Gordon equations, *Inverse Problems* **3, no.1** (1987), 37-49.

[21] M. Boiti, J. Leon, L. Martina, and F. Pempinelli, Scattering of localized solitons in the plane, *Phys. Lett. A* **132** (1988), 432-439.

[22] B. A. Borisov and S. A. Zykov, The dressing chain of discrete symmetries and proliferation of nonlinear equations, *Theor. Math. Phys.* **115, no.2** (1998), 199-214.

[23] A. Calini, A Note on a Bäcklund transformation for the continuous Heisenberg Model, *Phys. Lett. A* **203** (1995), 333-344.

[24] F. Calogero and A. Degasperis, *Spectral Transform and Solitons*, North-Holland Publishing Company, 1982.

[25] H.-H. Chen, General derivation of Bäcklund transformations from inverse scattering problems, *Phys. Rev. Lett.* **33** (1974), 925.

[26] J. Clairin, Sur les transformations de baecklund, *Ann. Sci. Ecole Norm. Sup.* **3, supplement** (1902), 1.

[27] A. Coley et al, *Bäcklund and Darboux Transformations, the Geometry of Solitons*, CRM Proceedings and Lecture Notes, 29, American Mathematical Society, Providence, RI, 2001.

[28] M. M. Crum, Associated Sturm-Liouville systems, *Quart. J. Math. Oxford* **6 no.2** (1955), 121-127.

[29] G. Darboux, Sur une proposition relative aux equation lineaires, *Compt. Rend.* **94** (1882), 1456.

[30] A. Davey and K. Stewartson, On three-dimensional packets of surface wave, *Proc. R. Soc. Lond.* **A338** (1974), 101.

[31] L. A. Dmitrieva, The higher-times approach to multisoliton solutions of the Harry Dym equation, *J. Phys. A* **26** (1993), 6005-6020.

[32] L. A. Dmitrieva, Finite-gap solutions of the Harry Dym equation, *Phys. Lett. A* **182** (1993), 65-70.

[33] E. Doktorov and V. Rothos, Homoclinic orbits for soliton equations solvable via the quadratic bundle, *Phys. Lett. A* **314 (1-2)** 2003, 59-67.

[34] L. D. Faddeev and L. A. Takhtajan, *Hamiltonian Methods in the Theory of Solitons*, Springer-Verlag, 1987.

[35] M. G. Forest, S. P. Sheu, and O. C. Wright, On the construction of orbits homoclinic to plane waves in integrable coupled nonlinear Schrödinger system, *Phys. Lett. A* **266, no.1** (2000), 24.

[36] E. I. Ganzha, An analogue of the Moutard transformation for the Goursat equation $\theta_{xy} = 2\sqrt{\lambda(x,y)\theta_x\theta_y}$, *Theor. Math. Phys.* **122, no.1** (2000), 50-57.

[37] T. I. Garagash, About the modification of the Painleve test for systems of nonlinear partial differential equations, *Theor. Math. Phys.* **100, no. 3** (1994), 367-376.

[38] C. S. Gardner, J. M. Greene, M. D. Kruskal, and R. M. Miura, Method for solving the Korteweg-de Vries equation, *Phys. Rev. Lett.* **19** (1967), 1095.

[39] E. Goursat, Sur une equation aux derivees partielles, *Bull. Soc. Math. France* **25** (1897), 36-48.

[40] E. Goursat, *Lecons sur L'integration des Equations aux Derivees Partielles du Second Ordre*, volume II, 1902.

[41] C. Gu, H. Hu, and Z. Zhou, *Darboux Transformations in Integrable Systems*, Springer, 2005.

[42] A. Hasegawa and Y. Kodama, *Solitons in Optical Communications*, Academic Press, San Diego, 1995.

[43] B. Herbst and M. Ablowitz, Numerically induced chaos in the nonlinear Schrödinger equation, *Phys. Rev. Lett.* **62** (1989), 2065-2068.

[44] W. Heremah, P.P. Banerjee, and M.R. Chatterjee, Derivation and implicit solution of the Harry Dym equation and its connections with the Korteweg-de Vries equation, *J. Phys. A* **22** (1989), 241.

[45] R. Hirota, Exact solution of the Korteweg-de Vries equation for multiple collisions of solitons, *Phys. Rev. Lett.* **27** (1971), 1192.

[46] R. Hirota, A new form of Bäcklund transformation and its relation to the inverse scattering problem, *Prog. Theor. Phys.* **52** (1974), 1498.

[47] Infeld, Hull T. E., The factorization method, *Rev. Mod. Phys.* **23** (1951), 21-68.

[48] M. N. Islam, *Ultrafast Fiber Switching Devices and Systems*, Cambridge University Press, New York, 1992.

[49] Y. Ishimori, Multi-vortex solutions of a two-dimensional nonlinear wave equation, *Prog. Theor. Phys.* **72** (1984), 3337.

[50] M. Kaashoek and A. Sakhnovich, Discrete skew selfadjoint canonical systems and the isotropic Heisenberg magnet model, *J. Funct. Anal.* **228** (2005), 207-233.

[51] S. I. Kiselev et al., Microwave oscillations of a nanomagnet driven by a spin-polarized current, *Nature* **425** (2003), 380.

[52] B. G. Konopelchenko, Induced surfaces and their integrable dynamics, *Stud. Appl. Math.* **96** (1996), 9-51.

[53] B. G. Konopelchenko and V. G. Dubrovsky, Inverse spectral transform for the modified Kadomtsev-Petviashvili equation, *Stud. Appl. Math.* **86** (1992), 219-268.

[54] B. G. Konopelechenko, and G. Landolfi, On classical string configuration, *Mod. Phys. Lett. A.* **12** (1996), 3161-3168.

[55] G. L. Lamb, Jr., Bäcklund tansformations for certain nonlinear evolution equations, *J. Math. Phys.* **15** (1974), 2157.

[56] Y. Lan, and Y. Li, On the dynamics of Navier-Stokes and Euler equations, *Submitted, available at: http://www.math.missouri.edu/~cli/Dynamics.pdf* (2008).

[57] S. Leble, M. Salle, and A. Yurov, Darboux transforms for Davey-Stewartson equations and solitons in multidimensions, *Inverse Problems* **4** (1992), 207-218.

[58] Y. Li, Bäcklund transformations for the equation $\partial^2 u/\partial x^1 \partial x^1 + \partial^2 u/\partial x^2 \partial x^2 = f(u)$, *Applied Mathematics and Mechanics* **10, no.2** (1989), 139.

[59] Y. Li, Auto-Bäcklund transformations for some evolution equations, *Advances in Applied Mathematics and Mechanics in China, Internat. Acad. Publ., Beijing* **4** (1992), 91.

[60] Y. Li, Bäcklund transformations and homoclinic structures for the NLS equation, *Phys. Lett. A* **163** (1992), 181.

[61] Y. Li, Smale horseshoes and symbolic dynamics in perturbed nonlinear Schrödinger equations, *J. Nonlinear Sciences* **9** (1999), 363.

[62] Y. Li, Bäcklund-Darboux transformations and Melnikov analysis for Davey-Stewartson II equations, *J. Nonlinear Sci.* **10** (2000), 103.

[63] Y. Li, A Lax pair for the two dimensional Euler equation. *J. Math. Phys.* **42, no.8** (2001), 3552.

[64] Y. Li, Singularly perturbed vector and scalar nonlinear Schrödinger equations with persistent homoclinic orbits, *Studies in Applied Math.* **109** (2002), 19.

[65] Y. Li, Homoclinic tubes in discrete nonlinear Schrödinger equation under Hamiltonian perturbations, *Nonlinear Dynamics* **31, no.4** (2003), 393.

[66] Y. Li, Chaos and shadowing lemma for autonomous systems of infinite dimensions, *J. Dynamics and Differential Equations* **15, no.4** (2003), 699.

[67] Y. Li, Chaos and shadowing around a homoclinic tube, *Abstract and Applied Analysis* **2003, no.16** (2003), 923.

[68] Y. Li, Homoclinic tubes and chaos in perturbed sine-Gordon equation, *Chaos, Solitons and Fractals* **20, no.4** (2004), 791.

[69] Y. Li, *Chaos in Partial Differential Equations*, International Press, Boston, 2004.

[70] Y. Li, Persistent homoclinic orbits for nonlinear Schrödinger equation under singular perturbation, *Dynamics of PDE* **1, no.1** (2004), 87.

[71] Y. Li, Existence of chaos for nonlinear Schrödinger equation under singular perturbation, *Dynamics of PDE* **1, no.2** (2004), 225.

[72] Y. Li, Melnikov analysis for singularly perturbed DSII equation, *Studies in Applied Math.* **114** (2005), 285.

[73] Y. Li, Chaos and shadowing around a heteroclinically tubular cycle with an application to sine-Gordon equation, *Studies in Applied Math.* **116** (2006), 145.

[74] Y. Li, Strange tori of the derivative nonlinear Schrödinger equation, *Letters in Mathematical Physics* **80, no.1** (2007), 83.

[75] Y. Li, et al., Persistent homoclinic orbits for a perturbed nonlinear Schrödinger equation, *Comm. Pure Appl. Math.* **XLIX** (1996), 1175.

[76] Y. Li and D. W. McLaughlin, Morse and Melnikov functions for NLS pdes, *Comm. Math. Phys.* **162** (1994), 175.

[77] Y. Li and D. W. McLaughlin, Homoclinic orbits and chaos in perturbed discrete NLS system, Part I. Homoclinic orbits, *J. Nonlinear Sciences* **7** (1997), 211.

[78] Y. Li and S. Wiggins, Homoclinic orbits and chaos in perturbed discrete NLS system, Part II. Symbolic dynamics, *J. Nonlinear Sciences* **7** (1997), 315.

[79] Y. Li and A. Yurov, Lax pairs and Darboux transformations for Euler equations, *Studies in Applied Mathematics* **111** (2003), 101.

[80] S. Lie, Begrundung einer invarianthentheorie der Beruhrungstranformationen, *Math. Ann.*, **VIII, Heft 3** (1874), 215.

[81] S. Lie, Zur Theorie de Flachen konstanter krummung III, *Arch. Math. og Naturvidenskab* **V, Heft 3** (1880), 282.

[82] S. Lie, Zur Theorie de Flachen konstanter krummung III, *Arch. Math. og Naturvidenskab* **V, Heft 3** (1880), 328.

[83] S. V. Manakov, On the theory of two-dimensional stationary self-focusing of electromagnetic waves, *Soviet Phys. JETP* **38, no.2** (1974), 248.

[84] V. B. Matveev and M. A. Salle, *Darboux Transformations and Solitons*, volume 5, Springer Series in Nonlinear Dynamics, 1991.

[85] V. K. Melnikov, On the stability of the center for time periodic perturbations, *Trans. Moscow Math. Soc.* **12** (1963), 1.

[86] C. R. Menyuk, Nonlinear pulse propagation in birefringent optical fibers, *IEEE J. Quantum Electron* **23, no.2** (1987), 174.

[87] C. R. Menyuk, Pulse propagation in an elliptically birefringent Kerr medium, *IEEE J. Quantum Electron* **25, no.12** (1989), 2674.

[88] A. V. Mikhailov, A. B. Shabat, and R. I. Yamilov, The symmetry approach to the classification of non-linear equations: Complete lists of integrable systems, *Russ. Math. Surv.* **42** (1987), 3.

[89] S. Milner, Square patterns and secondary instabilities in driven capillary waves, *Journal of Fluid Mechanics* **225** (1991), 81-100.

[90] B. Mityagin, Spectral expansions of one-dimensional periodic Dirac operators, *Dynamics of PDE* **1, no.2** (2004), 125.

[91] R. M. Miura, *Bäcklund Transformations, the Inverse Scattering Method, Solitons, and Their Applications*, volume 515, Lecture Notes in Math., Springer-Verlag, 1976.

[92] Th. Moutard, Note sur lesequations differentielles lineaires du second ordre, *C. R. Acad. Sci. Paris.* **80** (1875), 729-733.

[93] V. S. Novikov, Reflectionless potentials for the acoustic spectral problem, *JETP Lett.* **72** (2000), 223.

[94] T. Ozawa, Exact blow-up solutions to the Cauchy problem for the Davey-Stewartson systems, *Proc. R. Soc. Lond.* **436** (1992), 345.

[95] C. Rogers and W. Schief, *Bäcklund and Darboux Transformations, Geometry and Modern Applications in Soliton Theory*, Cambridge Texts in Applied Mathematics, Cambridge University Press, Cambridge, 2002.

[96] C. Rogers and W. Shadwick, *Bäcklund Transformations and Their Applications*, Academic Press, New York, 1982.

[97] A. B. Shabat, Third version of the dressing method, *Theor. Math. Phys.* **121, no.1** (1999), 165-176.

[98] W. F. Shadwick, The Bäcklund problem for the equation $\partial^2 z/\partial x^1 \partial x^2 = f(z)$, *J. Math. Phys.* **19, no.11** (1978), 2312.

[99] J. C. Slonczewski, Current-driven excitation of magnetic multilayers, *J. Magn. Mater* **159** (1996), L1.

[100] H. Steudel, The hierarchy of multi-soliton solutions of the derivative nonlinear Schrödinger equation, *J. Phys. A: Math. Gen.* **36** (2003), 1931.

[101] S. I. Svinolupov and R. I. Yamilov, Explicit Backlund transformations for multifield Schrodinger equations. Jordan generalizations of the Toda chain, *Theor. Math. Phys.* **98, no.2** (1994), 207-219.

[102] A. P. Veselov and A. B. Shabat, Dressing chains and spectral theory of the Schrodinger operator, *Functional Anal. Appl.* **27, no. 2** (1993), 81.

[103] H. D. Wahlquist and F. B. Estabrook, Prolongation structures of nonlinear evolution equations, *J. Math. Phys.* **16** (1975), 1.

[104] H. D. Wahlquist and F. B. Estabrook, Prolongation structures of nonlinear evolution equations II, *J. Math. Phys.* **17, no.7** (1976), 1293.

[105] G. B. Whitham, *Linear and Nonlinear Waves*, Wiley-Interscience Publication, New York, 1981.

[106] S. A. Wolf et al., Spintronics: A spin-based electronics vision for the future, *Science* **294** (2001), 1488.

[107] O. C. Wright and M. G. Forest, On the Bäcklund-gauge transformation and homoclinic orbits of a coupled nonlinear Schrödinger system, *Phys. D* **141, no.1-2** (2000), 104.

[108] A. V. Yurov, Darboux transformation for Dirac equations with $(1+1)$ potentials, *Phys. Lett. A* **225** (1997), 51-59.

[109] A.V. Yurov, Conjugate chains of discrete symmetries in $(1+2)$ nonlinear equations, *Theor. Math. Phys.* **119, no.3** (1999), 419-428.

[110] A. A. Yurova, A. V. Yurov, and M. Rudnev, Darboux transformation for classical acoustic spectral problem, *Internat. J. Math. Sci.* **49** (2003), 3123.

[111] V. E. Zakharov and A. B. Shabat, Exact theory of two-dimensional self-focusing and one-dimensional self-modulation of waves in nonlinear media, *Sov. Phys. JETP* **43, no.1** (1972), 62.

Index

About the Authors

Y. Charles Li works in the areas of chaos in partial differential equations, Navier-Stokes equations and turbulence, nano-technology, biological mathematics, and complex systems. His honors include a Guggenheim Fellowship, an AMS Centennial Fellowship, and the Princeton University Merit Prize. His published books include *Invariant Manifolds and Fibrations for Perturbed Nonlinear Schrödinger Equations, volume 128 of the Applied Mathematical Sciences series* (Springer-Verlag, 1997), and *Chaos in Partial Differential Equations* (International Press of Boston, 2004).

Artyom Yurov works in the area of mathematical physics, with expertise in applying Darboux transformations to various physical problems.

Made in the USA
Monee, IL
08 November 2026

68136719R20096